T0332413

Human Cold Stress

Human Cold Stress

Ken Parsons

CRC Press is an imprint of the
Taylor & Francis Group, an **informa** business

First edition published 2021
by CRC Press
6000 Broken Sound Parkway NW, Suite 300, Boca Raton, FL 33487-2742

and by CRC Press
2 Park Square, Milton Park, Abingdon, Oxon, OX14 4RN

Library of Congress Cataloging-in-Publication Data

Names: Parsons, K. C. (Kenneth C.), 1953- author.
Title: Human cold stress / Ken Parsons.
Description: First edition. | Boca Raton : CRC Press, 2021. | Includes
    bibliographical references and index.
Identifiers: LCCN 2021007956 (print) | LCCN 2021007957 (ebook) | ISBN
    9780367551995 (hardback) | ISBN 9780367552008 (paperback) | ISBN
    9781003092391 (ebook)
Subjects: LCSH: Cold--Physiological effect. | Human beings--Effect of
    climate on. | Human beings--Effect of environment on. | Human
    engineering. | Human comfort.
Classification: LCC QP82.2.C6 P37 2021 (print) | LCC QP82.2.C6 (ebook) |
    DDC 612/.014465--dc23
LC record available at https://lccn.loc.gov/2021007956
LC ebook record available at https://lccn.loc.gov/2021007957

ISBN-13: 978-0-367-55199-5 (hbk)
ISBN-13: 978-0-367-55200-8 (pbk)

Typeset in Times
by Deanta Global Publishing Services, Chennai, India

To

Jane, Ben, Anna, Nancy, Thomas
Sam, Mina, Hannah, Richard and Edward

Contents

Preface

Being cold is a miserable human condition and most people try to avoid it. However people often find cold Arctic conditions and up mountains exhilarating and refreshing if they are active and have sufficient clothing especially when the sun is shining. Also, cold air is perceived as fresh and of high quality so there is a general feeling of well-being.

A person may not feel cold but we describe the environment as cold because it has the potential to cause unacceptable cold strain on the body. The environment provides the cold stress and the human response is the cold strain that can lead to discomfort, severe incapacity, injury and death.

This book defines cold stress and how people respond to it. It describes how to assess a cold environment to predict when discomfort, hypothermia, shivering, frostbite and other consequences will occur. It also advises on what to do about it to prevent unacceptable outcomes. Over 30 years of teaching human responses to cold made me discover 'early on' that describing the effects of cold had academic but limited practical impact. You have to feel it. Practical sessions in cold chambers and water baths were the only way to demonstrate how undesirable it was.

In the winter in climates where the weather turns cold, there are many more deaths than in the summer, mainly among the elderly. In the UK there are around 50,000 more deaths from January to March than from October to December. Cold is likely to be a contributing factor along with influenza (other pandemics) and maybe psychological factors. This is so consistent that it can almost be regarded as a natural phenomenon and is why elderly people go south for the winter. Whether pandemics, such as caused by the 2020 spread of COVID-19, will be part of the natural order and how much it is exacerbated by cold stress, remains to be seen. I am reminded of the linear model of main effects of cold (C) and a new virus (V) plus interactions (CxV) yet to be investigated.

Chapter 1 provides a description of factors that make up human cold stress and strain, including wind chill and thermoregulatory responses. Chapter 2 describes the body heat equation and modes of heat transfer between the body and the cold environment. Chapter 3 provides a detailed description of human thermoregulation including physiological, adaptive and behavioural responses. Chapter 4 considers metabolic heat production including shivering and Chapters 5 and 6 provide a comprehensive coverage of the important topic of clothing from properties, to specification, selection and evaluation.

Chapters 7, 8 and 9 consider assessment methods and include wind chill, a calculation of clothing required in any cold environment and computer models that predict physiological responses to cold. The effects of cold on human performance are described in Chapter 10. Chapters 11 and 12 consider human response to diverse cold environments, including cold water and up mountains, and diverse populations including age, gender and people with disabilities. Cold injuries and health are considered in Chapter 13, and skin contact with cold surfaces is presented in Chapter 14. Chapter 15 considers the assessment of existing cold environments including how to measure cold stress and cold strain, with a case study of the assessment of a cold working environment.

The book includes references and a useful index that conclude a comprehensive and focused coverage that provides fundamental understanding and practical application in the area of human cold stress.

Ken Parsons
Loughborough
January 2021

Author Biography

 **Ken Parsons** is Emeritus Professor of Environmental Ergonomics at Loughborough University. He has spent over 30 years conducting laboratory and field research into human cold stress. He was born on January 20, 1953, in northeast England in a coastal village called Seaton Sluice. He was part of the British Schools Exploring Society expedition to Iceland in 1970. He graduated from Loughborough University in ergonomics in 1974, obtained a postgraduate certificate in education in mathematics with a distinction from Hughes Hall, Cambridge University in 1975 and was awarded a PhD in human response to vibration in 1980, from the Institute of Sound and Vibration Research, Southampton University. He founded the Human Thermal Environments Laboratory at Loughborough in 1981 and was awarded a certificate in management from the Open University in 1993. Ken became head of the Department of Human Sciences in 1996 covering research and teaching in ergonomics, psychology and human biology. He was Dean of Science from 2003 to 2009 and pro-vice chancellor for research from 2009 to 2012. He was chair of the United Kingdom Deans of Science from 2008 to 2010.

In 1992, he received the Ralph G. Nevins award from the American Society of Heating, Refrigerating and Air-Conditioning Engineers (ASHRAE) for 'significant accomplishments in the study of bioenvironmental engineering and its impact on human comfort and health'. The Human Thermal Environments laboratory was awarded the President's Medal of the Ergonomics Society in 2001. He is one of the co-authors of the British Occupational Hygiene Society publication on thermal environments and has contributed to the Chartered Institute of Building Services Engineers publications on thermal comfort as well as to the *ASHRAE Handbook: Fundamentals*.

He has been a fellow of the Institute of Ergonomics and Human Factors, the International Ergonomics Association and the Royal Society of Medicine. He was a registered European Ergonomist and an elected member to the council of the Ergonomics Society. He has been a scientific advisor to the Defence

Evaluation Research Agency and the Defence Clothing and Textile Agency and a member of the Defence Scientific Advisory Committee. He has been both secretary and chair of the thermal factors committee of the International Commission on Occupational Health (ICOH), chair of the CNRS advisory committee to the Laboratoire de Physiologie et Psychologie Environmentales in Strasbourg, France, and is a life member of the Indian Ergonomics Society. He was a visiting professor to Chalmers University in Sweden and is a member of the committee of the International Conference on Environmental Ergonomics. He was an advisor to the World Health Organization on heatwaves and a visiting professor to Chongqing University in China, where he was leading academic to the National Centre for International Research of Low Carbon and Green Buildings. He was scientific editor and co-editor in chief of the journal *Applied Ergonomics* for 33 years and is on the editorial boards of the journals *Industrial Health, Annals of Occupational Hygiene* and *Physiological Anthropology*.

He is co-founder of the United Kingdom Indoor Environments Group and a founding member of the UK Clothing Science Group, the European Society for Protective Clothing, the Network for Comfort and Energy Use in Buildings and the thermal factors scientific committee of the ICOH. He was chair of ISO TC 159 SC5 'Ergonomics of the Physical Environment' for over 20 years and is convenor to the ISO working group on integrated environments, chair of the British Standards Institution committee on the ergonomics of the physical environment and convenor of CEN TC 122 WG11, which is the European standards committee concerned with the ergonomics of the physical environment.

Human Cold Stress

1

COLD

Human cold stress is the continuous and dynamic combined interaction of the effects on a person of variables air temperature, radiant temperature, air velocity and humidity, as well as clothing and activity, that can result in a drop in body temperature. The variables are termed the six basic parameters as fixed representative values are often assumed. It is not determined by one (e.g. air temperature) or a sub-set (e.g. air temperature and wind) of those factors but the combined effect of all six. In water, water temperature and flow will add to cold stress.

The effects of the cold stress can lead to unacceptable cold strain in terms of discomfort, distraction, loss in capacity to perform tasks, reduced health, cold injury and death. This can be caused by lowered body temperature (hypothermia) but is often due to the strain on the body (particularly the heart and in vulnerable people with underlying medical conditions) due to physiological response as well as people simply 'giving up'.

Being cold is a miserable experience and should not be confused with being in a cold environment which can be exhilarating, stimulating and pleasant. A cold environment is one that has the potential to cause heat loss from the body leading to feeling cold, discomfort and lower body temperatures. Exercise and clothing can keep the body warm in a cold environment and provide thermal satisfaction and pleasure. Toner and McArdle (1988). Young (1988) cites Leblanc (1975) 'man in the cold is not necessarily a cold man'.

There are three main human responses to cold stress. These are to *increase heat production:* by exercise, increased muscle tone (thermogenesis) and shivering; *reduce heat loss:* using clothing and physiological responses that lower skin temperature; and, most importantly, *behavioural changes:* that reduce the level of cold stress (move to a warmer area, seek or create shelter), reduce

1

exposed body surface area (change posture), increase activity and clothing insulation and many more depending upon culture, customs and opportunities available. 'Adaptive opportunity' that 'allows' behavioural responses is an important characteristic of the environment and should be a fundamental part of environmental design. It could be regarded as a seventh basic 'parameter' and is an essential consideration in any modern assessment of human cold stress.

A final point about cold is that it has no meaning in physics other than to say that a body of lower temperature (less average kinetic energy) is often said to be colder than one of higher temperature. Being cold is a biological and psychological phenomenon not experienced by inanimate objects or fluids. A brass monkey does not feel cold despite the inconvenient result of lowering its temperature.

PSYCHO-PHYSIOLOGICAL THERMOREGULATION

Feeling cold and uncomfortable is a major drive for human response usually in terms of behaviour. Comfort (or discomfort) can be regarded as a significant 'controlled variable' in human thermoregulation (Parsons, 2020). When uncomfortable we respond in an attempt to become comfortable. When comfortable there is no drive for change and we are satisfied with the thermal environment. When heat loss is significant we sacrifice comfort for survival and control internal body temperature to an optimum level of around 37°C (people are homeotherms). This is part of homeostasis where humans control their internal environment to optimum levels. As body temperatures fall, physiological responses include vasoconstriction (preventing warm blood reaching the skin) to lower skin temperatures (reduce heat loss) particularly in the feet and hands with associated discomfort and drive for behavioural response and shivering, which increases heat production. The environment is dynamic and this psycho-physiological response is a continuous interaction between the body and the environment.

WATER

Water is essential to life (over 50% of the body is water and it is a significant part of our environment) and for the temperature range on Earth we experience water in its states as liquid, vapour and solid (ice, snow, etc.) and its conversion from one state to another. Water influences heat transfer from the body in each of these three states. Evaporation of water vapour (at a rate of the latent heat of vaporization (2.45×10^6 J kg^{-1} or 41W of sweat loss at 1g/min (McIntyre, 1980)), causes a chill when cold, and heat loss by breathing. Liquid water turns to water vapour during evaporation and also causes heat loss by conduction (particularly if the body is immersed in water – and also convection if the water is moving). Liquid water also reduces the insulation of clothing. Water is at its highest density at 4 °C and when water freezes (at 0 °C (or −0.5 °C in cells and −2 °C for sea water)) it releases the latent heat of fusion (3.33×10^5 J kg^{-1}). It also expands so that when water in cells freezes it damages the cell contents and causes irreversible damage to the cell wall (frostbite).

We should note that when water changes from vapour to liquid, in condensation, the latent heat of vaporization is released, just as it is absorbed from the body when liquid sweat changes from liquid to vapour. When water changes from solid ice to liquid the latent heat of fusion is absorbed and released when freezing from liquid to ice. It takes heat to break the molecular bonds when water changes form, so energy is required but there is no change in temperature. This is relevant to heat transfer for people in cold environments.

Water evaporated to vapour and transferred to the inside of impermeable clothing will condense on the inside surface of the clothing and heat will be conducted to the outside thereby releasing a small amount of heat from the body by sweating even when wearing impermeable clothing. The build-up of liquid water will reduce the insulation of the clothing. Collins (1983) and Golden and Tipton (2002) however note that to evaporate 1 litre of water a person would lose up to 675 W, whereas if they wore an impermeable outer-cover (wind break) even with wet clothing it would take only 35 W to heat water from 4°C to a comfortable skin temperature at 33°C in one hour.

A layer of warm water will exist around a warm body in cold water. This will rise as it is less dense than the cold water so heat will be lost by natural convection but it will give insulation to the body in still water when compared with moving water which will reduce or eliminate this layer of warmer water so that the body surface rapidly becomes that of water temperature. This is a familiar experience for those used to taking cool baths and explains why immersion in cold flowing water is more dangerous than immersion in cold still water with restricted body movement.

WIND

For a warm body in air there is also a temperature gradient between the skin and clothing surface of a person and the lower temperature in the cooler air of the environment. There is a layer of warmer air therefore around the body and as it is less dense than the cold air outside; it rises at a rate proportional to the temperature difference. This is called natural convection and it can be significant in a cold environment, particularly from the exposed skin of the head and hands. Air flows upwards across the surface of the body and there is a plume of warmer air above the person. Air movement across the body caused by body motion or wind (or both) interferes with the air layer such that the body surface tends towards environmental temperature and hence clothing and skin temperature become close to air temperature due to the forced convection of the wind. The skin then feels cold and the effect is called wind chill. If the body is wet then evaporation will cool the body even further and the chilling effect will be much greater.

FEELING COLD

There are separate hot and cold sensors across the skin of the body, more in some places than others (Gerrett et al, 2015). They are nerve endings (no specialised sensor has been identified) that respond to temperature and particularly the rate of change of temperature. Cold sensors have a firing frequency from 2 Hz at 15°C, peaking at 9 Hz at 30°C, reducing to 0 at 37°C with an 'erroneous' paradoxical discharge above 45°C (Dodt and Zotterman,1952). There is adaptation if temperatures do not change and temperature and rate of change of temperature provide a firing rate that is interpreted along with other factors to provide a feeling of cold. They can also imply wetness as the body does not have sensors that can detect wetness directly (Filingeri et al., 2014).

Fanger (1970) (see also ISO 7730, 2005) describes a Predicted Mean Vote (PMV) index that predicts the mean sensation of a large group of people on a scale +3, hot; +2, warm +1, slightly warm; 0, neutral; −1, slightly cool; −2, cool; −3, cold from an equation that includes values of the six basic parameters. For optimum (comfort = neutral) conditions the body must be in heat balance and skin temperatures (and sweat rates) must be at comfortable levels. That is for steady-state conditions. In changing environments (including rapid or transient changes) rate of change of skin temperature is important. Parsons (2020) emphasizes the importance of changes in air velocity as it influences the insulation provided by the air layer at the skin. So exposed skin temperature tends towards air temperature in wind.

Draughts cause a feeling of cold and can provide discomfort and a stiffness of muscles if prolonged. The percentage of people dissatisfied due to a draught (when they otherwise have neutral sensation) can be predicted as a

Draught Rating (Risk) DR = $(34-ta)(v-0,05)^{0.62}(0.37Tu+3.14)$ for air temperature ta (°C), v is local mean air velocity (ms⁻¹) and Tu is the local air turbulence, the standard deviation of the local air velocity divided by its mean (%), (ISO 7730, 2005; Olesen, 1985). Turbulence is significant as it reduces the insulation of the air layer more effectively than laminar air. When a person is already cold in terms of the whole body, the draught rating is expected to be more severe than when the person is otherwise comfortable as the draught rating merges into wind chill. A radiant 'draught' occurs when there is asymmetric radiation for example when sitting next to a cold window on a train (Underwood and Parsons, 2005).

Feeling cold is a consequence of physiological regulation as vasoconstriction 'attempts' to preserve heat but lowers skin temperature particularly in the hands and feet. The drive for behavioural response is very strong and taken as a whole provides a framework for the consideration of human cold stress.

Feelings of thermal discomfort and physiological responses can be predicted by a calculation based upon the heat transfer between the body and the environment. This is suggested as a first step in any assessment of human response to the environment and is termed the thermal audit (Parsons, 1992). Interpretation of the thermal audit can contribute to the assessment of human cold stress, prediction of thermal strain and provide a basis for environmental and system design. It is presented in the following chapters.

The Body Heat Equation

2

THE THERMAL AUDIT IN THE COLD

The first law of thermodynamics ensures that we can account for heat into and out of the body without loss and hence conduct a thermal audit (Parsons, 2014). The human disposition as a homeotherm is to attempt to maintain a 'constant' internal body temperature at around 37°C. For any object to maintain constant temperature the heat inputs to that object must balance with the heat outputs from the object. If a person has a net heat gain then their body temperature will rise and if a there is a net heat loss it will fall. For a constant temperature then the body must be in heat balance. If we derive equations that quantify avenues of heat inputs to the body and heat losses from the body we can set up a body heat equation and for constant temperature we can describe the body heat balance equation (see Figure 2.1).

An important starting point is that energy for the body is generated in the cells by the metabolism of food. The total amount of energy produced is termed the metabolic free energy production in Joules or metabolic rate (M) in Joules per second or Watts (often normalised to Watts per square metre of body surface area (Wm^{-2})). Much of that metabolic rate is produced as heat (H = M-W) which is transferred from the body to the environment. The remainder is used for mechanical work (W). Human thermoregulation can be considered to be about the relative preservation and dispersion of metabolic heat to control body temperature.

During cold we must preserve sufficient heat to maintain body temperature; however, when exercising in the cold, excess heat must be lost (through clothing) to the environment. This is driven by the differences between skin (and clothing) surface temperatures and air temperature as well as differences in radiant temperatures and the differences in water vapour concentration at the skin and in the environment (related to humidity). It is also influenced

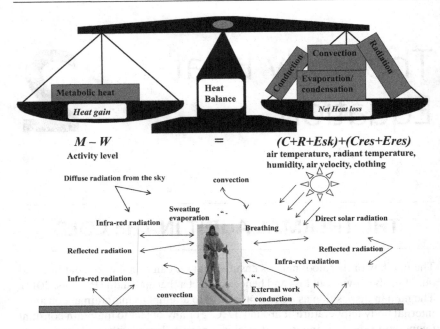

FIGURE 2.1 The body heat balance equation and avenues of heat transfer between the body and the cold environment.

by air velocity across the body and the thermal properties of clothing (e.g. insulation and vapour permeability). Any heat balance equation and any consideration of human response to cold must therefore consider air temperature, radiant temperature, air velocity, humidity, clothing and activity (to estimate metabolic heat production). These are termed the six basic parameters (see Chapters 4, 5 and 9).

The four mechanisms of heat transfer between the body and the environment are conduction (K), convection (C), radiation (R) and evaporation (E). The heat balance equation is $(M - W) + K + C + R + E = 0$ where by convention, M and W are heat gains and K, C, R and E are heat losses. A negative heat loss is a heat gain (including condensation for evaporation).

HEAT AND TEMPERATURE

Heat is a form of energy with 1 calorie defined as the heat required to raise 1g of water by 1°C. The SI unit is the Joule(J) where 1cal = 4.2 J. The specific heat capacity (c) of a substance is the heat to raise 1 kg of the substance by 1°C (=1K). For water c= 4.2 kJ kg⁻¹ K⁻¹. For the human body it is around 3.5 kJ kg⁻¹ K⁻¹. If a person weighs 100kg and our body heat calculations show a net input of 350 KJ then there will be an increase of 1°C in body temperature.

As a body gains heat it raises its average kinetic energy and this is termed its temperature. Absolute zero temperature is when the molecules in a body are stationary and it has no energy. This is zero degrees kelvin (0°K). Temperature scales are often defined in terms of values at critical points, so the Celsius scale has 0°C (273.15K) at the freezing point of water and 100°C (373.15K) at the boiling point of water (1°C temperature difference = 1 K temperature difference). The Fahrenheit scale is 0°F at the freezing point of a particular salt solution; 32°F at the freezing point of water and 212°F at the boiling point of water (giving 180 degrees between the freezing and boiling point of water). To convert °C to °F multiply by 9, divide by 5 and add 32. To convert °F to °C subtract 32, multiply by 5 and divide by 9. (Internal body temperature is around 37 °C = 98.6 °F; interestingly, a temperature of −40 °C =−40°F.)

The human body obeys the laws of thermodynamics. The first is that energy is neither created nor destroyed. The second is that there is a natural direction of transfer from hot to cold bodies and from high concentration to low concentration. From the perspective of human response to cold environments this means that we can account for (it will always exist) all avenues of heat loss and gain and calculate whether there is heat balance, positive heat storage (where the body temperature will rise) or negative heat storage (where the body temperature will fall). It also means that heat will be lost from a warm body to a cold environment and from a wet skin to a 'dry' cold environment where the water vapour concentration will be lower.

BODY HEAT TRANSFER BY CONDUCTION (K)

Conduction is the transfer of heat by direct 'contact', from molecules of a substance to adjacent molecules of lower energy such as in one direction from one (hotter) end of a material, sequentially to the other end at lower temperature. Militky et al. (2008) define it as the proportionality factor in the Fourier equation describing the steady-state one-directional transport of heat through a body of cross-sectional area A and length L due to thermal difference ΔT. The rate of heat transfer by conduction (K) is determined by the thermal conductivity of the material (Kc) and the temperature difference $(t_1 - t_2)$ over the distance transferred (d). $K = Kc\ A\ (t_1 - t_2)/d$ (W) (or $K = Kc\ \Delta T/L$ (Wm⁻²)).

Thermal conductivity of solid matter is around 1–5 (Wm^{-1} K) with water 0.6, ice 2.24 and air 0.024 Wm^{-1} K, respectively. So if cold weather clothing were solid, 0.1m thick, with a thermal conductivity of 0.1 W m^{-1} K^{-1} and skin was at a comfortable temperature of 33°C a person in cold air of 3°C would lose K = 0.1 (33–3)/0.1 = 30 W m^{-2} of heat by conduction, so for a person of surface area 2 m^2, 60 W.

As clothing is not solid, is usually dry and often contains air, any heat loss from the whole body surface by conduction is usually small. When it reaches the external environment, conduction to the surrounding air is negligible when compared with other avenues of heat loss and is often ignored when assessing people in cold environments.

Heat transfer by conduction from the body can often cause feelings of cold and even skin damage (ISO 13732 – 3, 2005) usually over specific areas of the body such as the hands and feet. This will be determined by skin temperature and can be predicted as a contact temperature between the skin and the cold surface (see Chapter 14).

Water has a thermal conductivity of 0.6 Wm^{-1} K; so immersion in water at 5 °C with a skin temperature of 33°C would give a heat loss by conduction of K = 0.6 × 2 × (33–5) = 33.6 W for a person with a surface area of 2 m^2. Conduction also occurs within the body into blood (thermal conductivity similar to that of water) and through tissues.

So the environmental parameters that influence heat transfer from the human body to a cold environment by conduction are the surface temperature of the body (skin or clothing) and the surface temperature and thermal properties (particularly thermal conductivity) of the surface in contact with the body.

BODY HEAT TRANSFER BY CONVECTION (C)

Heat loss by convection is due to movement of fluid (liquid or gas). For people in environments on Earth this is most commonly air and sometimes water. Fluid is heated (e.g. by the body), becomes less dense and rises to be replaced by more dense air (water, etc.) due to gravity. Heat is therefore transferred upwards and away from the body at a rate related to the difference in temperature between the body surface and the surrounding environment (air, water, etc.). Forced convection occurs when the fluid heated by the body is replaced by moving fluid in the environment (e.g. by wind or water currents).

Heat transfer by convection (C) can be a mixture of natural and forced convection and is a significant avenue of heat loss from a person to a cold environment. It can be estimated by the equation $C = hc\,(ts-ta)$, where ts is the surface temperature of the body and ta is air temperature (tw for water temperature etc.). The convective heat transfer coefficient (hc) is related to temperature differences for natural convection and air (water, etc.) and to air velocity for forced convection. Wind chill is often related to heat loss by forced convection and is considered in Chapter 7. Environmental parameters related to heat loss by convection in air are therefore air temperature and air velocity.

BODY HEAT TRANSFER BY RADIATION (R)

All objects emit radiation in the form of electromagnetic waves according to the formula $R = \varepsilon\,\sigma\,T^4\ Wm^{-2}$ where ε is the emissivity of the object (from 0 to 1. Lower values for reflective surfaces and 1 for a black body), $\sigma = 5.67. \times 10^{-8}$, is the Stefan-Boltzman constant and T is the temperature of the object in degrees Kelvin (K). The higher the temperature the shorter the peak wavelength of the emitted spectrum (Wein's law). Electromagnetic waves travel through a vacuum, and the sun at a surface temperature of around 5,800 K emits relatively short wavelength radiation when compared with indoor temperatures of around 293 K that emit longer wavelength radiation. This is important in human heat transfer as white surfaces (clothing, snow, etc.) will reflect short-wavelength solar radiation but indoors, out of the sun, with lower-temperature surfaces they will emit and absorb all wavelengths and behave like a black body. The colour of clothing is therefore important for thermoregulation in outdoor solar environments and the presence of snow or water will reflect solar radiation onto the body.

There are three methods of radiant heat transfer from the sun to a person. These are direct, diffuse (from the sky but not directly in line with the sun) and reflected radiation (see Figure 2.1). A maximum of around 1000 Wm^{-2} of direct solar radiation arrives on the earth on a bright sunny day. This is reduced by clouds, etc., in the atmosphere. It should also be remembered that just as hot surfaces transmit heat to the body by radiation, radiant heat is lost to cold surfaces.

There is a radiant heat exchange between a person and the environment such that there are gains and losses. The net radiant heat into the body can be represented by $R = hr\,(Ts^4 - Tr^4)$. Where Ts is the absolute temperature (K) of the surface of the body (mean skin temperature (Tsk) for a nude person and mean temperature of clothing (Tcl) when clothed) and Tr is the mean radiant temperature. The heat transfer coefficient by radiation (hr) includes the Stefan-Boltzman constant, the orientation of the body and radiation (see Underwood and Ward, 1966; Santee and Gonzalez, 1988; Parsons, 2014) and emissivity. Mean radiant temperature is therefore taken as an essential environmental parameter for the assessment of cold stress. The assessment of solar radiation can be included as part of mean radiant (or even plane radiant) temperature, but as its effects can be dominant, it is often considered as a separate component (Hodder and Parsons, 2007).

BODY HEAT TRANSFER BY EVAPORATION (E)

Heat is lost from the body to a cold environment by the evaporation of moisture and sweat from the skin and through breathing. When exercising in the cold the body sweats even though there is often capacity to lose sufficient heat by convection, and when exercise ceases it can cause a 'chill' and wet clothing.

Heat is transferred from the skin by evaporation according to the equation $E = he\ (Psk,s - Pa)\ Wm^{-2}$ where the evaporated sweat at the skin is assumed to be at saturated vapour pressure at skin temperature (Psk,s) and the partial vapour pressure in the air is Pa. The equation can be modified by adding the effects of skin wettedness (how wet the skin is) and the thermal properties of clothing (e.g. vapour permeability). The temperature will reduce away from the skin to clothing temperature (tcl) and be evaporated into the air so heat is lost from the skin to the environment. If the temperature falls to the dew point within the clothing then it will condense, releasing heat but turning to liquid. If it reduces to 0°C and below, it will freeze (and may be easy to remove). If sweat does not evaporate at all then liquid sweat may be removed by wicking into the clothing. These mechanisms are related to the thermal properties of clothing. The heat transfer coefficient by evaporation (he) is estimated using the heat transfer coefficient by convection (hc) and the Lewis relation (LR) from Lewis (1922) where $he = LR\ hc$ and $LR = 16.5\ K\ kPa^{-1}$.

For breathing in a cold environment heat will be lost by saturated vapour from the lungs, leaving the mouth at around 34°C and is often seen condensing as a vapour trail into cold air.

So heat transfer by evaporation from a person into a cold environment will be influenced particularly by environmental parameters such as air temperature, air velocity and humidity and will be restricted, often in a complex way, by the thermal properties of clothing.

Human Thermoregulation in the Cold

3

A SYSTEM OF PHYSIOLOGY AND BEHAVIOUR

All people have a conscious natural drive to be comfortable and seek pleasure, as well as a continuous unconscious drive to maintain an internal body temperature at around 37 °C for survival. They achieve this through behaviour and metabolic heat in a psycho-physiological control system of thermoregulation (Parsons, 2019). Much research has been conducted into adaptive behaviour and thermal comfort (de Dear et al, 2020) but adaptive behaviour is particularly important for survival in the cold (Young, 1988).

Thermoregulation occurs in the cells of the body as well as for the whole body. A cold cell will have an impaired function when compared with one at optimum temperature and if it freezes, it will suffer irreversible damage internally as well as to the cell wall. Each cell must be protected as well as the whole body. If internal body temperature as a whole falls, extremities such as the hands and feet become cold, organs may fail (particularly the heart) and a person can die.

Physiological thermoregulation involves methods to increase metabolic heat production as well as to preserve heat loss to the environment mainly by reducing heat (blood) flow to the skin and hence lowering skin temperature. Human responses to cold and related phenomena include: vasoconstriction, blood shunting, arterio-venus anastomoses, critical temperature, thermogenesis, shivering, brown fat, cold-induced vasodilatation (CIVD) and piloerection.

BLOOD FLOW AND DISTRIBUTION IN THE COLD

Blood carries nutrients and oxygen to the cells around the body and removes 'waste' products such as carbon dioxide. It also removes excess heat but this is not a waste product as it is an integral part of homeostasis and controlling internal body temperature. Heat is also transferred by conduction through tissues but most heat is transferred in the body by convective heat transfer in the blood.

When blood reaches the skin, heat is lost from the body to the environment when environmental temperature is lower than skin temperature. If it is desirable to lose heat from the body to the environment then it is optimum for as much blood as possible at body temperature to flow close to the skin surface; hence vasodilation. If it is desirable to preserve heat, as under cold stress, then vasoconstriction ensures that as little blood as necessary flows to the skin surface. There is therefore an active and continuous system of vaso-control which leads to the concepts of body core and shell.

Body core temperature (tcr – internal body temperature including the internal organs and brain) is 'defended' by varying body shell temperatures (outside tissues including the skin and often represented by mean skin temperature – tsk). This is highly influenced by blood flow and distribution. In cold stress, shell temperature is relatively low with maximum vasoconstriction and in heat stress it is relatively high with maximum vasodilation. These are controlled by the sympathetic nervous system and the neural transmitter norepinephrine (noradrenaline) (see Kenney et al., 2012 for a fuller discussion).

Core and shell are concepts and cannot be measured; so indicators such as, sub-lingual, rectal, tympanic or aural temperature are used for internal body temperature (core) and mean skin temperature for shell temperature. It is a consequence of blood flow in the skin, that to estimate mean skin temperature few measurements are needed in hot conditions (e.g. a weighted average of four points: chest, arm, thigh and calf (Ramanathan, 1964)) and a larger number is needed in cold conditions where there is a more heterogeneous distribution of temperatures across the skin (e.g the ISO 14 point method that includes hands and feet (ISO 9886, 2004)).

A weighted average of core and skin temperature provides the concept of mean body temperature (t_b) where $t_b = \propto tsk + (1 - \propto) tcr$. Where $\propto$ is the relative 'mass' of skin to core often taken as ranging from 0.1 for vasodilation to 0.3 for vasoconstriction. For a cold person with a core temperature of 36.5 °C and a mean skin temperature of 30 °C, the mean body temperature will be 34.55 °C. For a comfortable person at rest with a core temperature of

36.8 °C, mean skin temperature of 33 °C and an alpha of 0.15, the mean body temperature is 36.23 °C.

COUNTER-CURRENT HEAT EXCHANGE

A consequence of vasoconstriction is reduced local skin temperatures in the hands and feet causing discomfort and loss in dexterity. To avoid loss of heat, cooler blood returning from the skin is directed in veins that pass warm blood in arteries, cooling the blood in the arteries, warming the blood in the veins and returning the heat to the body 'core'. This is the counter-current heat exchange (in hot conditions blood returns to the skin to enhance heat loss).

ARTERIO-VENUS ANASTOMOSES

In a presentation on work in the cold, I saw a Scandinavian scientist 'flap' his hands to beat alternative arms to artificially demonstrate the mechanism of arterio-venus anastomoses. These are small blood vessels that prevent the blood from entering capillaries and hence away from the skin when cold. This is a short-cut system in the limbs (often called blood shunting) that avoids heat loss to the environment (and does not usually involve mechanical beating). The anastomoses open to promote heat loss and close to avoid it.

The totality of vaso-control is that 'core' temperature is defended by variations in 'shell' temperature, determined by blood flow and direction and hence regulating the heat transfer from the skin to the environment. When in heat balance, 'core' temperature is regulated within 1 °C (from the brain to the rectum) but 'shell' temperature can vary greatly under cold stress from warm head to cold toes. Areas of the body where the surface area to mass (or volume) ratio is large are particularly vulnerable to low skin temperatures (e.g., fingers, toes, nose and ears).

PILOERECTION

Piloerection is where 'hairs stand on end' which creates a layer of still air providing insulation next to the skin. As humans have relatively little hair it is often regarded as insignificant. The effects are however under-investigated and insulation due to pilo-erection may be significant under clothing, in still and moving air and for preserving heat generated by shivering.

SELECTIVE BRAIN BLOOD FLOW

Cabanac (1995) suggested that when a person is hot, cooler blood from the body (nose and skin of the face) flows to the brain (selective brain cooling – SBC) and when cold this is prevented. Some animals have a fine matrix of blood vessels (rete) that allow effective heat exchange near the brain but this is not found in humans. For people, this remains controversial but in practical terms, it has not been reported that a person survived due to selective brain cooling or heating through this mechanism.

HEAT PRODUCTION IN THE COLD

When a person is cold, vasoconstriction reduces blood flow to preserve heat but, if the temperature falls to a 'critical temperature', muscle tone increases and metabolic heat production is increased (thermogenesis). If temperatures continue to fall, muscles start to shiver which is an asynchronous contraction and relaxation of the muscle (10–12 Hz) often starting in the neck and upper back (Slonim, 1952) that produces heat but no work. This is involuntary shivering that is driven by the fall and rate of fall in body skin and core temperatures. Voluntary shivering (and muscle contraction) is a behavioural response often caused by rapid change in skin temperature. A further oscillation in response is a hunting-like phenomenon where shivering occurs, the body warms, shivering stops, the body cools, shivering starts and so on.

Shivering can increase resting metabolic rate by up to around six times for short durations and by about double for longer durations (Hayes, 1989; Collins, 1983). It is inhibited by exercise, some drugs (e.g. insulin), O_2/CO_2 levels, anaesthesia and some illnesses.

Brown fat is fat with cells that contain an abundance of mitochondria and infused with blood vessels (hence brown appearance) to allow rapid production and distribution of heat (Collins, 1983). It is often found in the upper back so conveniently placed to supply the brain. It is said to be effective in some animals (e.g. hedgehogs) and maybe in human infants but not adults. This has been an area of continuous debate. From a practical perspective there seem to have been no reports of human survival in the cold due to activation of brown fat.

Fat is highly influential in ensuring survival in cold water due to its insulative properties ("survival of the fattest", (Golden and Tipton, 2002)) the utilization of fat (and other resources) to produce heat depends upon the rate of oxygen disassociation and other metabolic processes, temperature and more. Despite sufficient fuel therefore, heat production may be insufficient for survival.

Some human responses are not directly related to thermoregulation. Eating food provides a small increase in heat production and is termed the specific dynamic action of food. It lasts for a number of hours but is insignificant when compared with the metabolic advantages provided by the food in producing metabolic heat. The diving reflex is a gasping response caused by a rapid change in skin temperature on the face usually as a consequence of rapid entry to cold water. It may be an attempt to take in oxygen but can lead to drowning. The Q_{10} relation describes how chemical reactions (hence metabolic processes and heat production) increase for every 10 °C increase in temperature. As temperature falls so the reverse will occur and often prolongs survival times in cold water for example due to reduced metabolic oxygen requirement and where cooling is rapid, particularly in children. Drugs and particularly alcohol affect the thermoregulatory system and are considered by Collins (1983) who also cites a study in Alaska that found that alcoholic stupour was the main factor in incidents of hypothermia (greater than aircraft accidents and shipwrecks). It is possible that cold may reduce the effects of drugs by slowing chemical reactions involving metabolic rate and drug uptake.

THE HUMAN THERMOREGULATION SYSTEM IN THE COLD

Hayes (1989) provides a detailed description of physiological protection of people in cold environments. He notes that overall people are tropical animals and equipped for responding to heat. Cold is undesirable and in general human response is to avoid it. Parsons (2019, 2020) suggests a psycho-physiological system of human thermo-regulation for people in hot, moderate and cold environments. A part of that system, showing responses to cold, is elaborated in Figure 3.1.

The system of psycho-physiological thermoregulation in cold environments is part of a more complete system for hot, neutral and cold environments (includes vasodilation, sweating, etc.). It is continuous and dynamic and

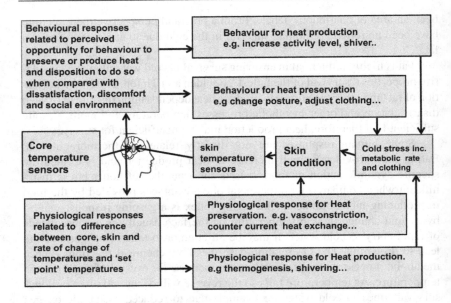

FIGURE 3.1   Psycho-physiological system of thermoregulation in the cold.

is always part of the function of a living person from birth to death. Although thermoregulation must take place at the level of the cell and is the integrated and interactive system across the body, it is represented as a cybernetic whole-body feedback system and a description can enter at any point.

To start with the cold stress 'box' (Figure 3.1), this is made up of the environment (assume in air, but equally valid in, and under, water, in space, on other planets, etc.). In air, this can be represented by an integration of the effects of air temperature, radiant temperature, humidity and air velocity. When combined with the effects of metabolic heat production and clothing insulation it provides the primary and essential components of human cold stress which directly influences skin temperature. When combined with blood flow, piloerection, moisture and sweating this will determine skin condition. Humans have no specialised sensors for wetness (implied from temperature and mechanical sensors) and for thermoregulation, 'cold' is determined by cold temperature sensors in the skin. The frequency of firing of the cold sensors is related to temperature and rate of change of temperature. The signal is transferred to the unconscious hypothalamus and the conscious cortex (feel cold) for interpretation and drive for action.

The hypothalamus also receives information from the internal body temperature sensors and when combined with that of integrated skin temperatures determines the output of this automatic, unconscious system that involves previous experience, adaptation, dead spaces and mistakes (interpret as hot at extremely low temperatures, etc.). The concept of a 'set point' (actually neither set nor a point) or 'set' desirable condition, has led to much debate (including what is actually controlled) but it is easy to show that people 'defend' a position where internal body temperature is around 37 °C (varies within 1 °C with time of day, time of month, position inside the body, exercise, fever and more).

In cold stress, there is a tendency for heat loss from a person and at internal body temperature below 'set point' a response is made in proportion to the difference between the sensed 'actual' internal body temperature and the 'set point' temperature. ASHRAE (2009) uses 36.8 °C for set point 'core' temperature for example.

There are two methods of physiological response, those for heat preservation and those for heat production. Heat preservation is mainly by control of blood flow mainly by vasoconstriction and to some extent by piloerection. Heat production is mainly by thermogenesis and shivering. Heat preservation mechanisms alter the skin condition, mainly reducing the skin temperature to reduce the gradient for heat loss to the environment. Heat production is considered as a reduction in the cold stress to which the person responds, hence completing the physiological cycle.

Cold stress and skin condition also affect the skin temperature sensor signals detected by the conscious cortex, providing a feeling of cold and a drive for a behavioural response. The greater the dissatisfaction, the greater the drive to do something about it (from dissatisfaction to satisfaction and comfort). This will depend upon what opportunities are available (perceived to be available) in the environment (adaptive opportunity). This also depends upon a threshold of inclination. It is often inconvenient to instigate actions such as altering heating controls or adding clothing insulation. The inconvenience may not be judged worth the perceived benefits. The social context may inhibit behaviour. A person may not want to draw attention to themselves or they may be nervous about being presumptive or upsetting others in a room, in a formal setting for example. The outcome is continuously under review and mainly conscious. It could be that there is no thermoregulatory behaviour because the threshold for action has not been reached in that context (or that there is no discomfort or dissatisfaction; Figure 3.1).

As for physiological thermoregulation, the two methods of thermoregulation using behaviour are those that promote heat production and those that promote heat preservation. Increased heat production is mainly by increased metabolic heat production through increased activity. Heat preservation is obtained through a multitude of methods often related to learned behaviour and culture. Increasing clothing insulation is the most obvious and often used method. Methods that reduce the cold stress include seeking shelter, avoiding wind and shadow, adjusting heating controls, closing windows, moving to a different location and more. The combination of behavioural methods for heat production and preservation influences the cold stress and completes the psychological part of the thermoregulation system. The conscious behavioural and unconscious physiological systems work in harmony to preserve or achieve internal body temperature and comfort (Brager et al, 2004; Parsons, 2020).

Human Metabolic Heat in the Cold

4

METABOLIC HEAT PRODUCTION

Each of the millions of living cells in the human body produce energy and the total is its metabolic rate. Most of the energy is produced in the mitochondria in the cell by 'burning' glucose in oxygen and most is produced as heat. If the metabolic heat production was not released into the environment then a person at rest would increase in temperature by around 1 °C per hour and they would die of hyperthermia after about four hours. Even in cold environments some metabolic heat must be released from the body.* The rest must be carefully managed and is vital for survival.

UNITS

One calorie is the heat to raise one gram of water by 1 °C and is 4.186 Joules (J). The rate of heat production of 1 Js^{-1} is 1 Watt (W). The heat to increase the temperature of 1 kg of the human body by 1 °C is on average about 3.49 kJ. A resting adult produces metabolic heat of about 1 W per kg of body mass (including fluids, etc.); that is, 3.6 kJ hr^{-1}. So enough heat for a body temperature rise of just over 1 °C per hour if none escaped to the environment.

---

* I am reminded of the fate of a passenger on an aeroplane who dressed in all of his clothes to save on luggage charges.

McIntyre (1980) suggests a body temperature rise of 1°C per hour will be achieved by a body heat storage of 38 Wm^{-2}. Basal metabolic rate is around 43 Wm^{-2}, so a rise of 1.1 °C for a totally insulated fully resting person.

Metabolic rate is the energy produced by the cells and different cells produce more or less energy depending upon the type of activity. The metabolic rate for a person can be represented correctly as the amount of heat on average for a given number (mass) of cells. All other factors being equal therefore, the more cells the more energy. For a person sitting at rest therefore, all other things being equal, at 1.5 W kg^{-1}; a 50 kg person will produce 75 W and a 100 kg person will produce 150 W. Note as temperature rise will be related to mass, this will produce the same rise in temperature in both people. It is therefore advantageous to specify standard metabolic rate values based solely on activity type and not the person conducting the task.

Standard, normalised values are clearly approximations as some people are more muscular than others and some perform tasks more efficiently than others and so on. Nonetheless we can use a standard value for a standard activity by normalising the values with individual characteristics. The most obvious is to use the average mass of a population and provide metabolic rate for a given activity in terms of W kg^{-1}. A preferred method used in the assessment of human response to the thermal environment is to normalise using skin surface area. This is because heat is transferred into and out of the body mainly through the skin, and this is consistent with the 'thermal audit' rational method of assessing human response to the environment (Parsons, 1992). A standard average person in this context is often taken as a 70 Kg man with a 1.8 m^2 surface area (65 kg and 1.6 m^2 often taken for the average woman). The surface area of a person is usually taken as $A_D = 0.202 \, W^{0.425} \, H^{0.725}$ where W is the person's weight in kg and H is the person's height in m. A_D is the DuBois surface area derived from measurements on a limited number of people (DuBois and DuBois, 1916). An easier to calculate method, and more consistent in terms of units, would be waist circumference × height (similar to a simple method for avoiding obesity where twice the waist circumference should not be greater than the height).

MEASUREMENT AND ESTIMATION OF METABOLIC HEAT PRODUCTION

The 'true' metabolic heat production is the combined heat that the cells are producing and hence is related to the number of cells and how much each cell is producing. Different cells produce different amounts of heat and during activity, thermogenesis and shivering, muscle cells increase their heat production and dominate the contribution to metabolic rate. In a cold environment, chemical reactions such as those involved in metabolic processes reduce in their speed of reaction according to a Q_{10} relationship. Q_{10} is the increase in reaction rate for each rise of 10 °C and is around a value of 2.5 for most biological reactions. That is, an increase in reaction rate of about 9.6% per °C increase in temperature (Hardy, 1979) and hence a corresponding reduced rate of reaction per °C drop in temperature. This is less important than it first seems as, in the cold, a cell at low temperature will produce very little heat so doubling it, or halving it, will have little contribution. Muscle strength and speed of nerve impulses also decrease with temperature as well as slowing of movement due to an increase in (synovial) fluid viscosity.

Measurements of metabolic heat are of two main types. Those that attempt to measure the heat output from a person and those that measure the oxygen consumption and make assumptions about combustion of food. All methods of measuring metabolic heat production require skill and expertise but even with careful experimental control they are prone to significant error (Parsons and Hamley, 1989).

Heat output from a person can be measured using whole-body calorimetry where all of the heat produced by a person is accounted for usually by enclosing the person in a specialist room, chamber or suit and measuring the change in air temperature and humidity or flowing water temperature. Hardy (1979) and Murgatroyd et al. (1993) provide details. Whole-body calorimetry is useful for measuring the metabolic heat production of resting people in a laboratory but not for other activities typical of practical application.

Indirect calorimetry is mainly concerned with measuring oxygen consumption and using the calorific value of food to estimate how much heat is produced. For example, if the difference between inspired and expired air has shown that V litres per second of oxygen has been 'taken in' by the body and we estimate from measurements (bomb calorimetry) that the energy equivalent by burning food in oxygen is 20600 J L^{-1} then a simple calculation estimates metabolic rate (M) as M = 20600 V $(0.2093 - O_e)$ Watts (McIntyre, 1980; Weir, 1949; Liddel, 1963). Where 0.2093 is the fraction of oxygen in inspired air and O_e is the fraction of oxygen in expired air. If O_e is 0.1693 and V = 0.1 Ls^{-1} then metabolic rate for that activity is M = 82.4 W.

The energy equivalent of burning food in oxygen in J L^{-1} O_2 varies from an estimated normal diet, 20600; fat, 19610; carbohydrate, 21120; protein, 19480; and alcohol 20330. The respiratory quotient (RQ) is a measure of 'combustion efficiency' and is the value of CO_2 'given off' divided by the value of O_2 consumed. Values are for fat (0.71); carbohydrate (1.0); protein (0.84); alcohol (0.67) and for a normal diet (0.85). Note that for protein some combustible material is released as nitrogen in urine and is therefore not used to produce metabolic energy. The metabolic rate can then be estimated as $M = 21168$ $(0.23RQ + 0.77)$ V $(0.2093 - O_e)$ W.

Measurement of metabolic rate by indirect calorimetry therefore requires a measurement or estimate of the rate of oxygen taken into the body (V) and the oxygen content of expired air for a person conducting the activity of interest. It also requires a measurement of CO_2 produced if RQ is to be taken into consideration (and nitrogen content of urine if protein is to be taken fully into account).

Systems of measurement may well interfere with the value being measured. They include full face masks or mouthpieces with valves (and nose clip). Breath-by-breath analysis can be made or a sampling system can be used. Hyperventilation must be avoided. A system of collecting expired air in a large bag was developed by Douglas (1911). Expired air was collected in a large rubber-lined twill bag (approximately 1 m² and >50 L) using a three-way mouthpiece and a nose clip. The collecting bag must have minimum resistance by being at mouth level and have short tubes and not be overfull, to avoid air resistance. The activity must be conducted over a period of time without, then with, the mouthpiece, before the valve to the bag is opened. A representative period of at least five minutes of collection should be made.

There are many precautions (Douglas, 1911; Murgatroyd et al, 1993, Parsons, 2019). Reports of standing, frozen Douglas bags when measuring the metabolic rate of military personnel in frozen regions are not surprising, with additional dangers of skin damage from contact with metal valves and the messy extraction of valves and nose clips with associated saliva. Less cumbersome methods such as breath-by-breath analysis also have practical problems and all in all, although indirect calorimetry is probably the best method available, it is by no means perfect. ISO 8996 (2004) suggests an accuracy of within 10% of actual values but emphasizes the importance of the accuracy of instruments and procedures. When taking all practical issues into account this may be optimistic.

Parsons (2019) suggests that the average metabolic rate for a group may be obtained if CO_2 production and O_2 depreciation are recorded in a closed room (lecture room, office, etc.); however, this has its own practical problems. In some extreme conditions, such as in cold environments, it is inconvenient or impractical to use Douglas bags or other collection devices. The relationship between heart rate (HR) and metabolic rate (by indirect calorimetry) can be found in a more convenient environment, for individuals, over a range of activities. This relationship can then be used for predicting metabolic rate from HR which is easier to measure for activities in extreme and complex environments such as those conducted in the cold.

A doubly labelled water method (Murgatroyd et al., 1993; Parsons, 2014) provides a method of estimating CO_2 production by tracking the deterioration of an isotope taken in a drink and monitored in urine (blood or saliva). This is related to energy production but requires assumptions and a number of days to obtain results so it is useful as an average estimate of energy expenditure over a week or more and hence for food requirements, but not for estimating heat production when conducting a a task.

The US Army determined equations for the prediction of metabolic rate for soldiers carrying loads and walking up hills of different gradients at different speeds (Givoni and Goldman, 1971; Pandolf el al., 1977). They used 105W as the value for resting metabolic rate. Multiple regression equations for specific load-carrying conditions have also been proposed by Garg et al. (1978); Morrisey and Liou (1984), Legg and Patemen (1984) Randle(1987) and Randle et al. (1989).

STANDARDIZED ESTIMATES OF METABOLIC RATE

It is a limitation in any analysis and assessment of human response to cold that although metabolic rate is an essential factor in determining human response its accuracy of measurement, especially for an individual, is probably less than 20% of its actual value. For extreme cold therefore, predictions of cold strain for individuals are limited and physiological measurements such as skin temperature (mean and local) and internal body temperature should be monitored. The use of standardized values for estimating metabolic rate may therefore not be highly accurate; however, they may be sufficient and be the best available and will provide a consistency in analysis.

Basal metabolic rate is the energy produced when a person is totally relaxed which includes 18 hours of fasting and lying down to minimise any heat production by postural muscles. This is usually measured at around 1.0 W per kg of body mass (total mass including fluids etc., which do not produce metabolic heat). Sitting at rest values are around 1.5 W per kg of body mass. For a 70-kg person that is 70 W (basal) and 105 W (resting) and for a person of 1.8 m² surface area it is 39 Wm⁻² and 58 Wm⁻², respectively. A value of 58.15 W m⁻² is the standard value taken as a unit of 1 met and is a conversion from earlier units of 50 kcal m⁻² hr⁻¹. Basal metabolic rate and resting values are probably better estimates of metabolic rate than those for higher-level activities as they may be taken under the more controlled conditions of whole-body calorimetry.

A note should be made here that metabolic rate, which is often referred to as total free energy production, is the total energy produced and that some of it is used as work (move limbs, flex muscles, breath, etc.). The energy that is not produced as heat is referred to as work (W) such that metabolic heat production H = M − W. This work (often referred to as external work or mechanical work) is regarded as negligible for many tasks, although it can be up to 25% of metabolic rate for efficient tasks such as cycling. We must be careful however, to distinguish between mechanical work produced and the work required for the body to function. When considering heat production and the human body it is the work required to move the limbs etc., that is relevant, relative to the heat when carrying out that work. W is therefore usually negligible and often taken as 0. It is considered here as it is traditionally included when considering heat production and methods of heat transfer to and from the body (the heat balance equation – see Chapters 3 and 8). It also clarifies the concept of negative work (e.g. going downstairs gives more 'return' than invested) which is not applicable in this context. In our consideration going downstairs requires work and produces heat.

Databases of typical metabolic rates have been produced mainly by using indirect calorimetry on people carrying out tasks (e.g. Durnin and Passmore, 1967; Spitzer and Hettinger, 1976). Indirect calorimetry can also be used to gain data for a specific population or workgroup. ISO 8996 (2004) provides standard methods and data for determining the metabolic rate.

ISO 8996, 2004: ERGONOMICS OF THE THERMAL ENVIRONMENT – DETERMINATION OF METABOLIC RATE

The measurement and estimation of metabolic rate for a given activity is of great importance, much studied and requires expertise and skill with potential errors and experimental limitations. In practice, for the assessment of human cold stress, it is therefore most useful to establish standard values that have validity (especially when comparing activities). ISO 8996 (2004) provides values for activities in general and ISO 11079 (2007) provides values for use in the assessment of cold stress.

ISO 8996 (2004) provides a comprehensive set of standard values and methods, that can be useful for estimating the metabolic heat production for specific activities and populations. In addition to a detailed description of indirect calorimetry, methods proposed include estimates from a general description by activity level (low, high, etc.); general description by occupation (bricklayer, office worker, etc.) a summation method (basal metabolic rate + arm work + posture, etc.); a heart rate method (for individuals); and by basic activities (sitting, walking, carrying etc.).

It is probable that the methods provided are either overly complex for practical applications (e.g. indirect calorimetry) or require more detail than is often available, especially for particular individuals and groups. A simple description of the level of activity may be sufficient for practical assessment and as good as is available. ISO 8996 (2004) gives metabolic rate values in W m^{-2} for resting (65); low (100) moderate (165); high (230) and very high (290) for a person of surface area 1.8 m^2. So the value in Watts can be derived by multiplying the values by 1.8. More recent versions of the standard will simply use Watts directly. ISO 11079 (2007) is specifically concerned with human response to cold environments and provides 'standardized' metabolic rates for different activities. A modified version is provided in Table 4.1.

TABLE 4.1 Metabolic rate values for categories of activity (Watts) (Modified from ISO 8996, 2004)

ACTIVITY LEVEL	METABOLIC RATE (W)	RANGE (W)	TYPES OF ACTIVITY
Resting	115	100–125	Sitting or standing at ease. Relaxed at rest.
Low	180	125–235	Light manual work. Hand and arm work. Bench work. Slow walking-level surface. Arm and leg work (driving).
Moderate	300	235–360	Walking in heavy clothing. Hammering. Heavy driving.
Medium			Paced-level walking. Ski-ing downhill, walking in snow, Pneumatic hammer. Pushing wheelbarrow. Plastering.
High	415	360–465	Intense arm and trunk work. Carrying heavy material. Shovelling. Sledgehammer work. Sawing. Pushing heavy loads. Fast-level walking.
Very high	520	>465	Very intense activity. Running or walking very quickly. Digging, level or uphill ski-ing, climbing stairs, ramp or ladder.
Very very high	720	500–1000	Very intense activity sustained without breaks. Emergency and rescue work at high intensity. Mountain rescue.

Note: 1. Minimum values for basal metabolic rate or sleeping are around 80 W. 2.Sports activities such as running or ski-ing can be up to 1000 W or more depending upon the level of performance.

METABOLIC HEAT IN MAMMALS

Mammals are homeotherms and many exist in cold climates where a wide variety of techniques are used to survive and are an integral part of their lifestyles including procreation and ensuring the next generation of the species. Hardy (1979) provides a description of methods some of which are summarised below and are relevant to human responses to cold stress.

Shivering is an involuntary tremor of skeletal muscle produced by impulses passing down somatic motor nerves to the muscle. The autonomic nervous system is not implicated. The ingestion of food, particularly protein, increases heat production (Specific Dynamic action – SDA). It begins to rise after 1 hour and remains above basal level for a number of hours, contributing to the maintenance of body temperature. Cold is a potent stimulus to thyroid function. It activates areas of the hypothalamus which stimulates the thyroid to release hormones that increase the metabolism of cells. The release of adrenaline, in an immediate response to cold, increases metabolic heat especially due to the release of blood glucose concentration. In certain mammals, brown fat is 'switched on' by impulses passing down the sympathetic nerves. It is also stimulated by noradrenaline release from the adrenal medulla. The sympathetic nervous system is a major influence on skin blood flow and sweating (Hardy, 1979).

Many survival methods are used by animals but a major one is the use of hair and fur covering the skin. Heat transfer by conduction provides insulation but in some animals heat transfer by convection and radiation are also harnessed. Sweating is rare but panting is common. For humans to survive the equivalent is the use of clothing or more correctly clothing systems.

A common question in the UK parliament during conflicts is why do our soldiers suffer from cold when a polar bear (or other Arctic dweller) can live perfectly well in the cold. The answer is that people have adapted to all of the climates of the earth whereas animals are well adapted to their own climates. Try running the polar bear uphill for long periods or placing it in the tropics. Metabolic heat due to activity can be preserved or lost to the environment to maintain body temperature, if the required amount of clothing is worn. This is considered in Chapters 5 and 6.

Clothing in the Cold

5

KEEP WARM AND 'LOOK GOOD'

Most people wear clothing for most of their lives. In hot climates they wear less clothing than in cold climates where they wear it to keep the body warm as well as to protect exposed skin and extremities such as the hands, feet, nose, ears, mouth and eyes. Outside of tropical regions, even with fire and shelter, people wear clothing to provide comfort and survive. They also want to look and feel good. Whatever the climate, appearance is a consideration.

Clothing has many functions including protection and fashion as well as design to provide practical utility (e.g. for holding tools, weapons and accessories) and for cultural reasons to cover modesty for example. Clark and Edholm (1985) list protection (from heat and cold; mechanical abrasion and blows; animals, plants and germs; air, wind, rain, snow, ice, fire; and contamination such as dirt, food, oils, bio-contaminants and chemicals) as well as providing comfort and effectiveness in performing tasks. Renbourn (1972) and Newburgh (1957) provide comprehensive coverage of the topic and Crockford (1991) provides a link between the extensive fashion industry and the vast amount of research into materials for protective and other work clothing driven by much research into military ensembles to meet the challenge of fighting and surviving in campaigns across the world, including those in extremely cold climates.

Siple (1945) provides an early and extensive summary of what was known about cold weather clothing in part of the studies in the Antarctic to develop the wind chill index (Siple and Passel, 1945). Goldman and Kampman (2007) provide a series of papers that summarise the findings of military research. Goldman proposes the four Fs of clothing. These are fashion, fit, feel and function. He emphasises the importance placed on fashion despite practical necessities. This chapter considers the important area of the thermal requirements for clothing for people under cold stress.

CLOTHING PROPERTIES

Clothing is material over the skin that provides a resistance to heat transfer between the human body and the environment. The skin varies in condition mainly in terms of temperature and wetness. The environment is usually determined (in air) by air temperature, radiant temperature, humidity and air velocity. Gravity also plays a role and the body also moves due to activity. All of these factors influence the transfer of heat into and out of the body and the properties of clothing influence that heat transfer which is made up of conduction, convection, radiation and evaporation/condensation. For many clothing ensembles the heat transfer between the body and the environment is complex and not fully understood with many dynamic features. Current methods of considering clothing properties combine this dynamic behaviour into three properties. These are thermal insulation, vapour transfer and ventilation.

THERMAL INSULATION

After much conjecture it is now accepted that the thermal insulation of clothing is mostly determined by the still air trapped within the clothing and not the type of material fibres used or its weight or density. Adding heavy clothing layers is not necessarily of advantage.

Thermal insulation is also termed thermal resistance and is the inverse of thermal conductance. It is a property that describes the resistance to heat transfer between the body and the environment. Conduction has units of Watts per square meter per °C temperature difference so thermal resistance has units of m^2 °C W^{-1}. Pierce and Reece (1946) described one Tog unit as 0.1 m^2°C W^{-1} and is the thermal resistance of 1 m^2 of material measured on a 1 square meter heated plate. A 10 Tog sleeping bag would therefore be made of material of thermal resistance 1.0 m^2 °C W^{-1}. Although useful, this is not the thermal resistance that would be provided to a person wearing clothing of that material.

The body is not flat and is more dynamic than a flat plate. Gagge et al (1941) proposed the Clo unit to describe the thermal resistance of clothing worn by people. It was originally proposed to explain to military leaders the amount of clothing needed for a range of activities in (particularly cold) climates. The Clo unit was given the value of 0.155 m^2 °C W^{-1} (0.18 °C Cal hr^{-1} m^{-2}) but importantly, the m^2 term refers to per square meter of the human body and not of the material. The Clo value for clothing is therefore the insulation provided to the whole body (clothed parts, of different thicknesses and types, and unclothed parts) as a representation of the thermal resistance or insulation provided to the whole body. A simple example of the thermal insulation of a naked person wearing a necktie demonstrates the difference between a Clo and a Tog. The necktie may be made of a material where 1 m² of the material has a thermal resistance of 0.1 m² °C W⁻¹ and hence 1 Tog. The thermal resistance of the necktie to the whole of the otherwise naked human body would, however, be about 0.05 Clo (0.008 m² °C W⁻¹)not much to keep the whole body warm.

In its current form, 1 Clo is said to be the insulation provided by a typical business suit. Typical values are for a nude person (0 Clo); T-shirt and shorts (0.4 Clo), light trousers and long sleeve shirt (0.6 Clo); business suit (1.0 Clo); business suit and overcoat (1.5 Clo); cold store clothing (2.0 Clo); heavy cold weather clothing (2.5 Clo). Values greater than 2.5 Clo are difficult to achieve as adding clothing layers provides diminishing return on thermal insulation partly because the surface area for heat exchange with the environment becomes large. Examples of the thermal properties of clothing ensembles are provided in Table 5.1.

If the skin or clothing surface is at a higher temperature than the environmental (air, water etc.) temperature then the environment will provide additional insulation (restriction to heat transfer) to a person. For a nude body in air this is given the symbol I_a and is the insulation of the air around the body. The intrinsic insulation of the clothing does not include this air layer and is given the symbol I_{cl}. As the insulation of the air around clothing decreases with the increase in surface area and hence thickness of clothing, the symbol used is f_{cl}, which is the ratio of the clothed surface area to the nude surface area. The total insulation is then given as $I_T = I_{cl} + I_a/f_{cl}$. For this model of clothing, I_{cl} is intrinsic to the clothing and not influenced by the environment. The I_a is influenced by the environment (especially air velocity). For still air for example the insulation provided by the 'air layer' is around 0.7 Clo. The I_a can be considered as the inverse of the convective heat transfer coefficient; however, strictly it should include radiation so that $I_a = 1/(h_c + h_r)$.

VAPOUR TRANSFER PROPERTIES OF CLOTHING

The skin is always moist to some extent, and with sweating it becomes wet and then cools by evaporation. If we assume a model of clothing that receives saturated vapour at skin temperature and allows it to transfer through clothing to the environment (driven by a difference in vapour concentration) then an

TABLE 5.1 Basic insulation (Icl) values of clothing ensembles. (Adapted from ISO 9220 (2007))

CLOTHING ENSEMBLE	CLO
1. Briefs, short sleeved shirt, fitted trousers, calf length socks, shoes	0.5
2. Underpants, shirt, fitted trousers, socks, shoes	0.6
3. Underpants, coveralls, socks, shoes	0.7
4. Underpants, shirt, overalls, socks, shoes	0.8
5. Underpants, shirt, trousers, smock, socks, shoes	0.9
6. Briefs, undershirt, underpants, shirt, overalls, calf length socks, shoes	1.0
7. Underpants, undershirt, shirt, trousers, jacket, vest, socks, shoes	1.1
8. Underpants, shirt, trousers, jacket, coverall, socks, shoes	1.3
9. Undershirt, underpants, insulated trousers, insulated jacket, socks, shoes	1.4
10. Briefs, T-shirt, shirt, fitted trousers, insulated coveralls, calf length socks, shoes	1.5
11. Underpants, undershirt, shirt, trousers, jacket, over-jacket, hat, gloves, socks, shoes	1.6
12. Underpants, undershirt, shirt, trousers, jacket, over-jacket, over-trousers, socks, shoes	1.9
13. Underpants, undershirt, shirt, trousers, jacket, over-jacket, over-trousers, socks, shoes, hat, gloves	2.0
14. Underpants, undershirt, insulated trousers, insulated jacket, over-jacket, over-trousers, socks, shoes	2.2
15. Underpants, undershirt, insulated trousers, insulated jacket, over-jacket, over-trousers, socks, shoes, hat, gloves	2.6
16. Arctic clothing systems	3–4.5
17. Sleeping bags	3–9

important property of clothing is how easily it allows this vapour and hence heat to transfer. This is mainly relevant to people in air, not to people immersed in water. Units for clothing vapour transfer are Watts per m² of body surface area per Pascal (difference between the vapour pressure at the skin and the vapour pressure in the environment). Vapour resistance is therefore given in m² kPa W⁻¹.

There have been a number of terms and indices of vapour permeability (e.g. i_m – the moisture permeability index of Woodcock, 1962). For a description, see ISO 9920 (2007); Goldman and Kampman (2007); and Parsons (2014). To be consistent with our model for thermal insulation, it is useful to use the model of resistance to vapour transfer (the evaporative resistance to clothing) as total evaporative resistance (Re,t) is the intrinsic evaporative resistance (Re,cl) plus the evaporative resistance of the air (Re,a). That is Re,t = Re,cl + Re,a/fcl. In their extensive work to determine the thermal properties of clothing McCullough et al. (1989) estimated an evaporative air layer resistance of 0.014 kPa m² W⁻¹.

VENTILATION, AIR PERMEABILITY AND RESULTANT INSULATION OF CLOTHING

Air can transfer directly from the skin to the environment through gaps between the skin and clothing, for example through the neck, sleeves and cuffs as well as vents designed into the clothing. This is related to clothing design and fit, and can be increased by human movement that causes a pumping or bellows effect. Some air can also penetrate clothing and this is called air permeability. Early attempts at quantifying the effects of ventilation and pumping suggested a reduction in intrinsic insulation of up to 20% due to ventilation effects. Some attempts have been more specific and use empirical data involving human subjects, tracer gas methods or moving manikins to determine ventilation effects usually involving walking (Bouskill et al, 2002). A simple method could relate the loss of insulation to the level of metabolic rate, but this is limited as it is the interaction of human movement and clothing design, not the energy produced, that is the relevant parameter.

ISO 9920 (2007) and ISO 11079 (2007) define the parameter resultant clothing insulation as the actual insulation provided to clothing (including the effects of human movement and wind). Resultant (dynamic) total and basic insulation as well as resultant total and basic evaporative resistance are also described. Multiple regression equations involving wind and human movement are provided to correct basic and total insulation and evaporative resistance values to resultant values (see also Havenith et al, 2002; Parsons, 2014). ISO 9920 (2007) also provides a discussion of body posture, chairs, pressure and wet clothing.

Clark and Edholm (1985) demonstrate the dynamic characteristics of the 'layer' of air around the body using Schlieren photography. A detailed description is given for a naked, standing man in still air with natural convection, from laminar flow at the feet to the turbulent 1m or more plume of air above the head and moving upwards across the body at around 0.5 ms⁻¹. There are stagnation points and the layer of air thickens as it rises, providing insulation for the head and influencing the flow that removes carbon dioxide (and other constituents) from exhaled breath around the mouth and nose (much reduced by face masks). Forced convection due to external air movement changes and usually reduces the insulation provided by this dynamic air layer. Posture greatly affects the flow profile and clothing interferes with this pattern, is influenced by clothing design and can lose heat directly from the torso with a chimney effect at the neck. Trapping air in stagnation points is useful in cold environments, such as using long sleeves and cuffs pointing downward to trap warm air for insulation of the hands.

Although the ventilation of clothing is related to clothing material and clothing design, the actual heat transfer is by convection and evaporation, often directly from the skin to the environment, and is more correctly represented as a direct heat loss rather than a reduction in clothing insulation.

A simple model would be a transfer of air at the skin to be replaced by air from the environment. Fanger (1970) uses a similar model for estimating the heat loss from the body by breathing, related to metabolic rate. As a first approximation, identical equations for calculating heat loss from ventilation in clothing may be reasonable (See Chapter 8).

Latent heat loss from the skin can be represented as $E = V (Wsk - Wa) \lambda$, where V is the clothing ventilation rate (kg s^{-1}), Wsk is the humidity ratio of the 'expired' air, Wa is the humidity ratio of the 'inspired' air, both in kg water/ kg dry air, and λ is the latent heat of vaporization of water at skin temperature (2407 kJ kg^{-1} at 35 °C). If saturated air at 35 °C (Wsk = .037) is replaced at 1 litre per second (V = 1.225×10^{-3} kg s^{-1}) by environmental air at 5 °C and at 50% relative humidity (Wa = .003), then E = $1.225 \times 10^{-3} \times (0.037-0.003) \times 2407 = 0.100$ KJ s^{-1} = 100 W of heat lost to the environment for perfect exchange of saturated microclimate air inside of the clothing with environmental air from outside. For a relatively dry skin at 20% relative humidity, and skin temperature of 35 °C (Wsk = 0.007) then 0.0118 KJs^{-1} = 11.8 W of heat will be lost to the environment.

If we consider 'dry' heat loss by convection and the ventilation rate (V) 1 ls^{-1} (= 0.001m^{3}s^{-1}) V $\times$ cp (the specific heat of dry air at constant pressure) $\times$ the difference between skin temperature and replacement environmental air temperature (tsk – ta) provides an approximation. The specific heat capacity of air is about 1 kJ/kg·K. Density is around 1.3 kg/m^3 so volumetric heat capacity is 1.3 kJ m^{-3} °C^{-1}. For the above example, this gives a value of 0.001 m^{3}s^{-1} $\times$ 1.3 kJ m^{-3} °C^{-1} $\times$ (35 – 5) °C = 0.039 kW = 39 W. A total of vapour and dry heat transfer of 100 + 39 = 139 W.

If we consider enthalpy as the total of heat and mass transfer, then the loss of enthalpy from saturated air at 35 °C to air at 50 % relative humidity and 5 °C is 143 – 29.5 = 113.5 KJ kg^{-1}. As 1 L s^{-1} of air is equivalent to 1.225×10^{-3} kg s^{-1} so a loss of 1 Ls^{-1} of air from clothing will lose 139 W. For a sweating person most of the heat loss due to ventilation will be from the evaporation of sweat and for dry skin it will be less than around 30 W for each Ls^{-1} of ventilation due to imperfect mixing. A psychrometric chart provides information required to make the calculations (e.g. Parsons, 2014).

MEASUREMENT OF CLOTHING PROPERTIES

The two main properties of clothing, of interest to the assessment of human cold stress, are the dry insulation of the clothing (mostly due to air trapped in clothing) and the vapour permeation (evaporative resistance) of the clothing, especially if the person is active and sweating. These are the clothing properties that provide basic measurements and are often used in clothing specification. Wind brakes, vents and openings, air permeability, wicking, layers, attention to exposed skin and extremities and other factors are of importance and should be influential in clothing design but a starting point for measurement is to determine or specify the thermal insulation and vapour permeation properties of clothing.

The development of clothing can start with an analysis of the person under cold stress and a specification of the overall heat preservation required (Holmer, 1984). To select materials for clothing, heated flat plates (dry to determine thermal insulation and wetted for vapour transfer (ISO 11092, 1993) properties (wicking, etc.)) are useful to provide an indication of what would be appropriate. For specifying the heat transfer characteristics of clothing, however, the use of thermal manikins provides an indication of clothing properties (and importantly standardised methods) and the use of human subjects performing tasks in conditions of interest allows estimates of properties. Bouskill et al (2002) describe a tracer gas technique to quantify the ventilation properties of clothing on human subjects.

THERMAL MANIKINS

A heated thermal manikin is a human shaped 'dummy' that measures the heat transfer from an outer skin (usually heated to be kept at a constant temperature – often 34 °C) through clothing and air layer, usually in a thermal chamber at constant, standard conditions (ASTM, 2020).

The heat required to maintain the manikin surface at a constant temperature is a measure of the heat lost to the environment. By running the manikin nude (some time is usually required to reach a steady state) the insulation of the environment (Ia) can be measured (the reciprical of Watts per m² required per °C difference between skin temperature and air temperature). The manikin is then dressed in the clothing of interest and, for the same conditions, the heat required to maintain the surface of the manikin at the constant temperature is equal to the heat lost to the environment when clothed. This provides the total insulation (It) of the clothing under those conditions.

The intrinsic insulation (Icl – independent of the environmental conditions) is then Icl = It – Ia/fcl. The ratio of the surface area of clothing to the nude surface area (fcl = Acl/A_D) is the clothing area factor and is estimated for the manikin and clothing using photographic or scanning techniques, paper, string, etc. Estimations of the fcl value provide a limitation to the method and a standard equation used is fcl = 1 + 0.28 Icl, where the Icl value is in Clo (McCullough et al, 1985; ISO 9920, 2007). Other equations include the percentage of body surface area covered by the clothing. All equations are approximations, however, and undermine the accuracy of using a sophisticated thermal manikin.

Some researchers have suggested ignoring fcl and using the term effective clothing insulation, Icl_{eff} = It – Ia. This, however, avoids the issue and also undermines the accuracy of the method. 'Sweating', thermal manikins can be used with wet skin surfaces and when combined with dry insulation measurements allow estimates of the vapour permeation properties of clothing. Moving, usually walking, manikins can allow estimation of the bellows effects and clothing ventilation. Systematic measurements can therefore determine clothing properties that can be interpreted for practical application.

HUMAN SUBJECTS

Human subjects can be used to estimate the thermal properties of clothing and have the advantage of providing real human responses. These methods include the use of a human subject in a similar way to a manikin as well as user tests and trials in thermal conditions of interest. Zhou et al (2002) used a clothing ensemble with sewn in pockets containing heat transfer transducers to determine the heat transfer characteristics of clothed human subjects.

THERMAL MODELS

Models of human thermoregulation (Nishi and Gagge, 1977; Parsons, 2014; Parsons, 2019) can be used to estimate clothing properties by determining what clothing parameters would give identical results to the average of a group of subjects exposed to climatic conditions of interest. To determine ventilation properties, Parsons (1991) described a user test where a comparison was made between the mean skin temperature at the end of a one hour exposure to a cool environment while standing still and the mean skin temperature after a regime of large slow arm and leg movements to maximise ventilation effects. Using a two node model of thermoregulation (Nishi and Gagge, 1977), the effect of ventilation on clothing insulation is estimated. For all of the methods using thermal models, however, metabolic heat is an important parameter for which it is difficult to obtain an accurate estimate and any error will influence the estimate of clothing insulation.

Selection of Clothing for Cold Environments

6

CLOTHING FOR THE WHOLE BODY

To determine clothing that will preserve comfort and survival in cold environments, and allow tasks to be performed, we can make a rough calculation of how much thermal resistance is required for the whole body, consider specific requirements to ensure that extremities are protected and conduct tests and trials to see if it works. If possible we should 'design in' adaptive opportunity so that the clothing can be adjusted, to allow variation of thermal resistance with activity level, for example.

The calculation of the thermal resistance or insulation required (IREQ) can be made using the body heat balance equation (Gagge et al, 1941; Siple, 1945; Burton and Edholm, 1955; Holmer, 1984), or using a thermal model (e.g. Nishi and Gagge, 1977). ISO 11079 (2007) provides a method of calculating IREQ from the temperature difference across the clothing (difference between the mean skin temperature and the mean temperature of clothing) divided by the metabolic rate minus any heat loss by breathing and evaporation (which is balanced by any heat transfer by convection and radiation to the environment).

So if skin temperature is 33 °C and the air temperature is 1 °C in the wind (so the mean temperature of clothing is around 3 °C) the temperature difference across the clothing is 30 °C. If the metabolic heat is 120 W m^{-2} and the heat loss by breathing and evaporation is 20 W m^{-2} (so heat loss by convection and radiation is 100 W m^{-2}). Then IREQ = 30/100 = 0.3 m^2 °C W^{-1}. That is 0.30/0.155 = 1.94 Clo.

This is the required basic resultant clothing insulation (Icl,r). Appropriate clothing can then be determined by selecting clothing ensembles with basic clothing insulation values (Icl) that when corrected for ventilation and other practical considerations will give a value of 1.94 clo. For example, an ensemble of undershirt, underpants, insulated trousers, insulated jacket, over-trousers, over-jacket, socks and shoes has been measured on a thermal manikin as having a basic insulation value of Icl = 2.2 clo (see Table 5.1, ensemble 14). If we allow for ventilation and other practical effects, it is reasonable to assume that the resultant basic insulation (Icl,r) will be close to the IREQ value of 1.94 clo and at least provide a starting point for design. Other ensembles of similar clothing insulation may be a better starting point depending upon the requirements of the task as well as requirements to protect extremities on hands, feet and head.

PROTECTION OF THE HANDS

The hands, particularly the fingers and thumbs, are susceptible to heat loss as they have a large surface area for heat exchange when compared with their mass (volume). Covering the whole hand with mittens (or withdrawing the hand into a sleeve or pocket) reduces available surface area and preserves heat but generally reduces manual dexterity. Parsons and Egerton (1985) demonstrated this with one material and 'all' logical combinations of fingers and thumbs. An opposing thumb in a mitten configuration provides some dexterity and preserves more heat than when each digit is covered separately, providing some insulation but replacing the air layer around the finger and increasing the surface area for heat loss.

If each finger is considered as an exposed cylinder, then heat will be lost from the finger up to a total critical radius of finger plus insulation of $r_c = k/h$, where k is the thermal conductivity of the insulation material and h is the insulation of the air. Goldman (1994) provide insulation for the 5th (little) finger on the bare hand as 0.27 m^2 °C W^{-1} and so thermal conductivity of air as 3.7 W m^{-2} °C $^{-1}$, and if insulation of material is 0.1 m^2 °C W^{-1} (1 Tog) so the critical radius before insulation preserves heat is rc = 0.1/3.7 = .027 m or 2.7 cm. This demonstrates the principle but in practice the insulation of the air will be greatly reduced by air and hand movement and increased by any still dry air trapped in the glove. Determination of the insulation of gloves is provided in EN 511 (2006) but evaluation using human subjects is recommended particularly, as they will greatly influence the ability to perform some tasks.

Goldman et al. (1994) also provide times for the 5th finger to freeze, based upon wind chill and the thermal insulation of gloves. The general conclusion is that in extreme cold, gloves, unless actively heated, will not keep the hands warm. Even when heated, gloved hands in pockets are most effective.

Insulation will ameliorate the effects of contact with cold surfaces (particularly important for metals) and will provide some manual dexterity when working in the cold. There are, however, many glove/mitten configurations that can be matched to practical applications. Examples are bare hands or gloves with covering mittens that 'peel back' to allow tasks to be performed and gloves with bare thumb and fore-finger to handle ropes (pull triggers, etc.). An additional factor is that performing manual dexterity tasks increases hand temperature.

Wet gloves and hands should be avoided. Sweat and moisture accumulate alarmingly quickly in impervious gloves that may be essential to protect from toxic hazards. Liquid water will greatly reduce insulation of material (provided mainly by still air pockets). Pre-warmed or heated gloves provide advantage. Hand protection is usually traded with task requirements and manual dexterity, and the selection of glove and mitten configurations is part of clothing system design. For thermal comfort, survival and the avoidance of cold injuries, it is essential to pay particular attention to the protection of the hands from cold.

Siple (1945) provides the following practical advice (adapted and summarized) for protection of the hands in the cold:

1. Avoid lengthy exposure of the hands.
2. Do not touch surfaces, particularly metal, with unprotected hands. Use low conductivity material such as wool next to the skin with a windproof cover.
3. Consider using gloves with exposed fingers and avoid restricting circulation.
4. Use loose-fitting mittens and gloves.
5. Avoid using gloves alone in extremely cold conditions.
6. Avoid overdressing and tight clothing that restricts circulation.
7. Keep hands and hand-gear dry at all times or be able to change.
8. Be able to remove mittens and allow drying.
9. Insulate tools and enlarge them to make them easy to use.
10. Avoid loose cuffs to avoid trapping of snow.
11. Place frozen flesh next to warm skin (thighs, armpits etc.). Do not rub with snow.
12. Use pockets and do not lose clothing, especially mittens.

Further discussion of gloves for use in the cold is provided by Santee et al. (2017), Santee and Berglund (2001), Parsons (2014) and Goldman and Kampman (2007).

There is some debate about heating local areas of the body as they may send disproportionate signals to the whole-body thermoregulatory system, stimulating inappropriate response (e.g. vasodilatation in the cold). Further research is required probably on a case-by-case basis. This is considered below along with the use of active clothing.

PROTECTION OF THE FEET

Bare feet will feel cold at skin temperatures below around 15 °C and can be injured when cold and wet and when frozen. Vasoconstriction in the feet and the large surface area when compared to the mass (volume) available for heat exchange reduces the temperature of the foot. Protection of the feet is often regarded as the most important requirement "More than any other part of the body, the feet need special attention in cold climates" Siple (1945). It is particularly important to keep the feet dry. Modern materials have provided insulation and breathability at reduced mass than previously available when animal furs were used. There is debate about the effectiveness of new materials, both passive and active, and evaluation is usually required. For example, a permeable-breathable layer of material compressed by the foot between layers of other material is unlikely to have any significant effect. Siple (1945) provides a summary of the extensive experience gained in World War 2. This is summarised in the following points:

1. The feet need special attention in cold climates
2. Standing, walking and running produce heat and stimulate circulation, sitting and lying do not.
3. Wet feet can lead to a medical emergency
4. In extreme cold, freezing point occurs within the boot and sock system.
5. In moderate cold keep water out. Use large waterproof boots with absorbent socks and shoes. Large, snug, roomy cold climate foot gear, 2–3 sizes greater than usual, is essential
6. Avoid compression of the legs and feet that may restrict circulation.
7. Insensible perspiration can be significant. Avoid overheating of the body. Too much clothing can stimulate sweating and should be avoided.
8. Dry or change boots and socks (and overall clothing)
9. Loss of sensation indicates that the feet have frozen and should be warmed next to warm skin (e.g. warm torso of others). Do not rub with snow.

Failure to protect feet in the cold can lead to 'non-freezing' cold injury (often called trench foot – which has halted armies) and frostbite, which causes irrecoverable damage.

PROTECTION OF THE HEAD

The blood circulation in the head is generally maintained in all conditions and hence head temperature remains high and has a high potential for heat loss, particularly in the cold and when wet. It has a high surface area for heat exchange when compared to its mass (volume), particularly in the ears and nose.

Hair provides good insulation and has a significant effect in protecting against the cold and parts of the head are 'used to' cold exposure and resistant to it. Hats or other head covers are essential if no hair is 'available'. Siple (1945) suggests that ears; back of the neck; temples and forehead; throat; top of the head; chin; nose; cheeks; mouth; eyes and face all need special attention. For protection of the head:

1. Head bands or ear muffs are essential even in moderately cold weather as "ears are susceptible to quick and painful freezing".
2. A hat or cap should protect the back of the neck. Wear a headband to protect the head, including the temples and protect the throat.
3. The top of the head will be protected by hair until extremely cold when additional protection will be required.
4. The chin requires protection (caps with chin straps) in wind at very low temperatures. Beards can freeze and become unhygienic
5. Nose and cheeks need protection in blizzards and very cold weather using face masks that allow effective sight and breathing.
6. The mouth and eyes are required for sight and breathing and are the last to be protected with the exception of bright sunlight or reflected light conditions where it is of highest priority to avoid 'snow-blindness'. Wear effective sunglasses and avoid metal components that can make contact with the face.
7. Freezing of the face is rapid and often causes a sharp twinge of pain. It spreads across the face and is best detected by a 'buddy'

THE DESIGN AND EVALUATION OF CLOTHING FOR COLD WORK

Consideration should always first be given to whether it is necessary for people to be exposed to cold and ways to avoid it (e.g. heated fork-lift truck cabs), before specialist clothing is used. Design of clothing should recognise that it is an integral part of the work (or other) system and 'job' and that workplace design, as well as human characteristics, activity and task analysis, is a primary and holistic consideration. Potter et al (2020) provide a cold weather ensemble decision aid (CoWEDA) for selecting clothing for work in the cold. Reinertsen (1998) considers test methods for protective clothing and off-shore workers from Norway.

The three 'D's for PPE (personal protective equipment) of donning, doffing and disposal (cleaning, storage, etc.) are important (actually, wearing is also important). The thermal requirements for clothing can be determined from the human heat balance equation and physiological criteria for acceptable thermal strain (see chapter 8). The calculated required clothing insulation (IREQ) can then be used to select from clothing ensembles with appropriate thermal insulation as a starting point for design (see Table 5.1). Identification of further requirements will lead to the inclusion of pockets and belts, and specific consideration should be given to extremities such as hands, feet and ears. Vapour permeation and ventilation will also be important for active and varied work and where sweat may accumulate.

Prototype (or existing) clothing ensembles can be evaluated using (representative) human subjects in user performance tests, trials and surveys (including diaries). User performance tests are conducted (often in a thermal chamber) representative of the conditions for which the clothing has been designed. User performance trials are conducted by workers under actual working conditions where some control is lost but realism is preserved. Survey and diary methods provide information over longer periods and evaluate the clothing system while it is being used in the application.

Dependant variables (measures) to indicate clothing performance include physiological measures (heart rate; blood pressure; internal body temperature; mean skin temperature; local skin temperature, sweat rate); subjective measures (sensation and comfort over areas of the body and overall, and acceptability and satisfaction); behavioural measures (distraction due to cold; attention off task; adjusting clothing; using pockets, moving away from cold environment); and performance measures (test measures such as pin or peg placing; performance at actual tasks).

An integration of measures compared with set criteria for acceptability in the context of interest will provide the clothing evaluation. The concept of clothing within a window of application (an envelope of climates for which the clothing is applicable) was pursued by Bethea and Parsons (1998), inspired by the UK scientist Geoff Crockford. It provides a practical method of clothing specification but the idea has not been 'taken up'.

'SMART' AND ACTIVE CLOTHING

I often considered clothing worn by hypothesised futuristic space travellers (e.g. as in the TV series 'Star-Trek') as the usual selected stretch fabric that authors naively thought would be worn. The lack of style and fashion alone would seem to negate this conception. If the clothing was smart clothing, however, it may be active to ensure optimum heat control, comfortable skin temperature, and eliminate discomfort caused by sweat as well as providing all sorts of other services. Smart clothing to date is at an early stage of development and is mainly used to preserve or generate heat or evacuate sweat.

From the perspective of humans in cold environments active and smart clothing includes at least five methods that can be used for influencing the whole-body and local thermal environments. Heated clothing (often vests) usually uses wires or conducting fabric and an electric current to heat the fabric. Faulkner et al (2013) proposed heated shorts for sports cyclists. Gels that provide exothermic reactions are also available. Heating of the hands using heated gloves is common especially where manual dexterity needs to be preserved and power supplies are available such as for pilots in cockpits. Battery power is also becoming sufficient for limited application although it will be reduced if batteries become cold. In a user trial in a large freezer room I observed that heated gloves were most effective when placed in pockets when the hands were not in use. Heated pockets may well have been a more convenient solution. Heated seats in vehicles can also preserve comfort, even at low air temperatures (Brooks and Parsons, 1999) and provide thermal comfort solutions for electric vehicles of the future.

Phase change materials are materials incorporated with small capsules (now possible in manufacturing processes) that change state, usually from solid to liquid or vice versa, supplying or withdrawing latent heat when required. Paraffin (with problems of flammability) is often used and if set at say 34 °C, below which it is solid and above which it is liquid, latent heat will be supplied when the body becomes cool and withdrawn as it becomes hot. In user tests I found that the amount of heat involved was an order of magnitude less than that required for whole-body ensembles; however, some change in sensation was detected when used in gloves.

Memory materials retain a previously set shape when they reach a set temperature. If that shape were large and spacious when above the temperature and flat below it, then if set at 35 °C when the skin starts to become hot and sweaty, air gaps and ventilation will be promoted and below that temperature a flatter shape may promote insulation.

Tubed undergarments and vests have been used for pilots and in space suits as well as for divers. Flat tubes against the skin may provide advantage (not 'proven') and water or other liquids can be pumped at the required temperature into the suit. The temperature range to avoid skin damage will be narrower for liquids than for air so care must be taken. Their application is also limited as pumping systems with associated power requirements will be needed.

Nano-machines are a category of a range of tiny 'machines, incorporated into materials, potentially matched to the clothing requirements. Micro pumps could pump air and liquids away from the skin to the environment. Most nano-machines currently claim to alter the permeability of materials, providing water proofing for example when required. Micro-chips incorporated into clothing may also have wider use in the future.

My involvement with developing and evaluating clothing, mainly in the context of human thermal requirements, provided the observation that new materials and ideas seem to be proposed very regularly and claim to provide the 'breakthrough' solutions that will enhance performance at sport, and other activities. These seldom turn out to be instant breakthroughs and although progress has been made it is systematic. There is no substitute for the evaluation of clothing ensembles in the systems and context within which they will be used.

Whether the clothing involves smart clothing or not, it is essential to integrate any garments with the clothing system and identify its objectives. An important point about smart and active clothing is that local effects may influence the effectiveness of thermoregulation and stimulate heat loss. For example, heating the hands may send erroneous signals of whole-body comfort or even heat stress to the hypothalamus that will stimulate mechanisms of heat loss that are not desirable when in the cold and attempting to preserve heat.

When using smart and active clothing systems it should be remembered that weight and generation or restriction of the evaporation of sweat may cause problems. In an evaluation of liquid cooling in a neck scarf, sweating was restricted so any benefit had to overcome the loss in effective thermoregulation as well as providing additional cooling (Bouskill and Parsons, 1996). For heated gloves therefore reduced vasoconstriction, which is a very effective mechanism for avoiding heat loss, will have to be traded off with the vasodilation caused by warming the hands. It is clear therefore that whatever the design of clothing there is no substitute for evaluation of the whole clothing ensemble and system.

Wind Chill 7

WIND AND LOW TEMPERATURES

Safe temperatures for activity in the cold based upon air temperatures alone will not be valid unless wind is taken into account. Radiation and humidity as well as clothing, activity level and weather conditions, will all have influence but practical experience shows that wind requires special consideration.

Wind chill can be considered to be the effect of wind on bare skin that creates a chilling sensation. Air at 35 °C blown across dry skin at a comfortable 33 °C will not be described as wind chill. If the skin is wet or the air is at a significantly lower temperature than the skin temperature then cooling will occur that will be described as a chill. Wet hair on the head is often considered as the cause of a 'chill' and, as every mother knows, should be avoided in cold environments. A draught could be considered as a special case of wind chill in the region of thermal discomfort. As well as the effect of wind on cooling exposed skin, wind chill could be generalized to include the effect of wind on the whole body even if clothed.

Logically, wind chill is the contribution of wind (air velocity) to the cold strain when a person is exposed to an environment that is defined by air temperature, air velocity, humidity, radiant temperature, clothing and activity. In many contexts, radiant temperature, humidity, clothing and activity level are defined leaving air temperature and air velocity to provide an indication of wind chill. It also allows a trade-off to be made between air temperature and air velocity such that the air temperature that would give the same effect on a person if the air were still can be calculated (the chilling temperature). This allows statements from weather forecasters such as 'the air temperature is minus five degrees outside but, if we take wind into account, it will feel like minus 20 degrees'.

An 'equivalent chilling temperature' could then be defined as 'the temperature of a cold environment that would give equivalent cold strain to a person in still air as in the actual environment defined by actual air temperature, radiant temperature, air velocity, humidity, clothing and activity. A wind chill index would then be a single number that indicates the effects of wind chill on the body, providing scale values related to cool and cold sensations and to physiological effects such as exposed skin freezing.

THE WIND CHILL INDEX

One of the earliest and certainly the most widely used wind chill indices was developed by Siple and Passel (1945) from work in Little America, Antarctica, carried out in 1941 by the United States Antarctic Service. They placed cylinders of water (one eighth of an inch thick pyrolin, approximately 15cm long and 6cm diameter; rumoured to be large empty bean cans) on poles in a range of positions and monitored the time taken to freeze 250 g of water. The time (t) from a cessation in decrease in temperature, initial freezing at 0 °C, to final freezing at 0 °C, was measured.

The latent heat of fusion is 79.71 calories per gram (334 J g^{-1}) and 1cc of water weighs 1 gram. A square meter of frozen water 1 cm thick (10,000cc) would liberate 797,100 gram calories per square meter = 797.1 kg cal m^{-2} in cooling time t in hours and given the symbol Ko in kg cal m^{-2} hr^{-1}. This was taken as the measured rate of total cooling. Still air measurements were made in an enclosed pit and all other measurements were made in a range of environmental conditions, in freezing temperatures and darkness such that radiant temperatures were less than air temperatures and along with very low humidity were considered insignificant.

The effect of cooling was determined by men in warm clothing facing the wind. Time to feel a 'sharp twinge of pain' indicating flesh freezing, was taken. Around twenty separate subjects experienced pain on the nose, eyelids, cheeks, wrist, side of temple and chin.

The kata cooling time of Hill (1919) and an equation involving the square root of the air velocity from Winslow et al (1936) were investigated and an improved equation was derived for the wind chill index (WCI). This is WCI = $(10v^{1/2} + 10.45 - v)(33 - ta)$, where v is the air velocity (ms^{-1}) and ta is the air temperature (°C). WCI is in kg cal m^{-2} hr^{-1} (or × 1.16 for Wm^{-2}). For effects less severe than freezing skin (cold, cool, etc.), subjective reports of people on treks in the Antarctic were used (Gold, 1935 and others). A standard scale was adopted that ranged from 0 (kg cal m^{-2} hr^{-1}) for calm conditions in the dark at 33 °C to 2500 towards an absolute limit of tolerance to cold.

Siple and Passel (1945) explain that an estimate of cooling power can be made using estimated skin temperature, but for a standard value and potential cooling the value of 33 °C is used in the equation. This was the comfort mean skin temperature that Gagge et al (1941) had used for the definition of the clo value. Siple and Passel (1945) were aware that a comprehensive wind chill index would involve radiation and humidity as well as clothing and activity, and factors related to weather (snow); however, they concluded that corrections and additions to their formula could be made and that the WCI was the total cooling power of the atmosphere in complete shade without regard to evaporation (total dry-shade cooling).

The WCI equation of Siple and Passel (1945) can be used to calculate the equivalent chilling temperature (tch), which is the temperature that would provide the same chilling effects as the actual environment but only if the air was calm ($v = 1.8$ ms^{-1}). Re-arranging their equation for $v = 1.8$ ms^{-1} and using the calculated WCI value for the actual conditions (same effect) provides the equation tch = 33 − WCI/22 °C (or 33 - WCI/25.5 °C for WCI in Wm^{-2}). An example would be in air temperature of −20 °C and air velocity of 9 ms^{-1}, then WCI = 1667 kg cal m^{-2} hr^{-1} so tch = 33 − 1667/22 = −43 °C. 'It may be minus twenty degrees outside but with the wind it will feel like it is minus 43 degrees'.

Despite recognition that any wind chill index would include all six basic parameters of air temperature, air velocity, humidity, radiant temperature, clothing and activity, the WCI of Siple and Passel (1945) has been used in military and industrial applications. For Antarctic conditions a scale of WCI values (in kg cal m^{-2} hr^{-1}) suggests that at WCI = 1200, conditions for travel start to become unpleasant. At WCI values around 1400 freezing of exposed flesh begins and this is often taken as a limit for work. Around WCI = 2000 conditions are considered dangerous, and exposed flesh freezes within 1 minute, and WCI = 2500 is intolerable and to be avoided (it is towards the limit of cooling in mid-winter Antarctica).

THE WIND CHILL TEMPERATURE (t_{WC})

ISO 11079 (2007) for the assessment of cold stress, proposes a general cold stress index for the whole body (IREQ) and a wind chill temperature that is a wind chill index based upon the cooling of the face (Osczevski and Bluestein, 2001). The equation was first adopted by Environment Canada and the US National Weather Service in 2001, who produced 'wind chill equivalent temperature charts', and was validated using human subjects at the Defence and Civil Institute for Environmental Medicine, Toronto, Canada. The index is given the equation $W = 13.12 + 0.6215ta - 11.37V_{10}^{0.16} + 0.3965ta \times V_{10}^{0.16}$, where W is the wind chill index (also termed the Wind Chill Temperature - ISO use the symbol 't_{wc}') in °C; ta is air temperature (°C) and V_{10} is the 'meteorological' air velocity (kmhr^{-1}) measured on top of a 10m pole and is approximated by $V_{10} = v \times 1.5$ for air velocity (v) at 1.5m height (Note that 10 kmhr^{-1} = 16 mph = 2.8 ms^{-1}).

ISO 11079 (2007) defines the Wind Chill Temperature (t_{wc}) as "the ambient temperature, which at a wind speed of 4.2 kmhr^{-1} produces the same cooling power (sensation) as the actual environmental conditions". The standard provides four categories of risk: (1) Uncomfortably cold (t_{wc} between −10 to −24 °C); (2) Very cold. Risk of skin freezing (t_{wc} between −25 to −34 °C); (3) Bitterly cold. Exposed skin may freeze in 10 minutes (t_{wc} between −35 to −59 °C); and (4) Extremely cold. Exposed skin may freeze within 2 minutes. (t_{wc} −60 °C and colder). (For further details, see ISO 11079, 2007; Shitzer and Tikuisis, 2012; and Parsons, 2014).

The 'new wind chill index' is becoming adopted world-wide and for industrial application by the influential American Society of Governmental Industrial Hygienists (ACGIH), who provide a chart for the 'Wind Chill Temperature Index' (ACGIH, 2021), which is similar to t_{wc} values provided in ISO 11079 (2007).

Environment Canada (2017) provide a table of guidance for all individuals (not just workers in good health) based upon the 'new wind chill temperatures'. These are summarized in Table 7.1.

WIND CHILL AND THE BODY MADE OF CYLINDERS

Parsons (2014) suggested that if WCI is to be restricted to dry conditions with no radiation component then the heat loss by convection (C) could be taken as the wind chill index as $C = 10\ v^{\frac{1}{2}}$ (tsk-ta) Wm^{-2} for the whole-body and $C = 20\ v^{\frac{1}{2}}$ (tsk-ta) Wm^{-2} for extremities (mean skin temperature (tsk) and air temperature (ta) in °C and air velocity (v) in ms^{-1}).

A further development for a simple wind chill index for practical use and based upon the principles of heat transfer, is to consider the whole body and its parts as cylinders of appropriate dimensions. A logical 'characteristic dimension' related to heat loss is to use the surface area to volume ratio (K) such that $C = K\ v^{\frac{1}{2}}$ (tsk-ta) Wm^{-2}. For an (open) cylinder $K = 2/r$, where r is the radius (m) of the cylinder ($2\pi rL\ /\ \pi r^2 L = 2/r$).

TABLE 7.1 Wind Chill Hazards and What To Do. Adapted from: "Wind Chill Index" Environment Canada (2017)

WIND CHILL TEMPERATURE (°C)	EXPOSURE RISK	HEALTH CONCERNS	WHAT TO DO
0 to -9	Low	Slight increase in discomfort	Dress warmly, stay dry
-10 to -27	Moderate	Uncomfortable. Risk of hypothermia or frostbite outside if inadequate protection.	Dress in layers of warm clothing with a wind-resistant outer layer. Wear hat, mittens; scarf, insulated waterproof footwear. Stay dry, stay active.
-28 to -39	High	High risk of frostnip and frostbite. Check face and extremities for numbness or whiteness. High risk of hypothermia if outside for long periods without adequate clothing or shelter from wind or cold.	Dress in layers of warm clothing with a wind-resistant outer layer. Cover exposed skin. Wear hat, mittens or insulated gloves, scarf, necktube or face mask, insulated waterproof footwear. Stay dry, stay active.
-40 to -47	Very high	Very high risk of frostbite. Check face and extremities for numbness or whiteness. Very high risk of hypothermia if outside for long periods without adequate clothing or shelter from wind or cold. Exposed skin can freeze in 5–10 minutes or less in high winds.	Dress in layers of warm clothing with a wind-resistant outer layer. Cover exposed skin. Wear hat, mittens or insulated gloves, scarf, necktube or face mask, insulated waterproof footwear. Stay dry, stay active.
-48 to -54	Severe	Severe risk of frostbite. Check face and extremities for numbness or whiteness. Very high risk of hypothermia if outside for long periods without adequate clothing or shelter from wind or cold. Exposed skin can freeze in 3–5 minutes or less in high winds.	Dress in layers of very warm clothing with a wind-resistant outer layer. Cover exposed skin. Wear hat, mittens or insulated gloves, scarf, necktube or face mask, insulated waterproof footwear. Stay dry, stay active. Be careful.
-55 and colder	Extreme	Danger. Hazardous outdoor conditions.	Stay indoors

If we assume that the surface area of the body can be approximated by the waist circumference × height, then for a cylinder of height 1.8m and circumference (waist) 1m (1.8 m^2 surface area), the radius is $1/2\pi$ (=0.159)m so the surface area to volume ratio K = 4π = 12.6. If K is the characteristic dimension (surface area/volume) then for a person of surface area 2m^2, Height 2m, K = 12.6; for surface area 1.5m^2, Height 1.0m, K = 8.4. A finger can be represented by a small cylinder, ears by half cylinders, nose by half a cone and so on. The actual heat loss and hence wind chill will be greatly influenced by local factors such as 'posture' and proximity of other body areas.

The heat loss from dry skin by convection is given by C = hc (tsk – ta), where tsk is skin temperature and ta is air temperature. The convective heat transfer coefficient (hc) for a cylinder in cross air flow is given by Dannielson (1998) as hc = 4.47 $d^{-0.38}$ $v^{0.62}$. Kerslake (1972), Fanger (1970) and others suggest that for the whole-body and forced convection, this is a function of the square root of the air velocity. A simplified equation is therefore C = K $v^{1/2}$ (tsk-ta). A simplification of the above equation of Danielsson (1998) is hc = 5 $d^{-0.38}$ $v^{0.5}$ so K = 5 $d^{-0.38}$, with estimates of K = 7.9 for the whole-body; K = 12.0, the head; K = 15.6, the hands; and K = 28.7, the fingers. This can be compared with the estimated surface area to volume ratios of K = 12.6 for the whole body; K = 30 (for a sphere SA/Vol = 3/r), the head; K = 50, the hands; and K = 419, the fingers.

This logical approach to calculating heat loss by convection is confounded by a number of practical factors such as surface area exposed, skin temperature and direction of wind. A practical approach is to consider each area of the body exposed. For the whole body a simple and valid approach would be to assume a value of K = 10 and for parts of the body values between 10 and 20 would be appropriate.

In summary a wind chill index can be proposed based upon a simplification of the heat balance equation for a person. For heat transfer by convection only (for extremities this is the spirit and concept of the wind chill index) WCIconv = $Kv^{0.5}$ (tsk – ta), where K is a constant related to the ratio of the surface area to the volume of a body part, v is air velocity, tsk is skin temperature (33 °C for comfortable temperature) and ta is air temperature.

Siple and Passel (1945) proposed the WCI in terms of convective heat transfer but noted that refinements could be made for other than dark dry conditions. If we add radiation (e.g. from the sun) then WCI_{R+C} = ho (33 – to), where ho = hc + hr and to is the operative temperature (= (hcta + hrtr)/(hc + hr)). If we approximate to with tar = (ta + tr)/2 then a simple WCI_{R+C} = $Kv^{0.5}$ (33 – tar). Further refinements may consider wind direction, and for solar or directional radiation, tr may be replaced by the plane radiant temperature (tpr).

A final refinement would be to consider wet skin (not really in the spirit of the WCI as in cold conditions wet skin is a crisis and should be avoided).

Then $E = LR \, w \, Kv^{0.5}$ (Psk,s − Pa), where LR is the Lewis relation (LR = he/hc = 16.5 K/kPa); Psk,s is the saturated vapour pressure at skin temperature (= 5.65 KPa at 33°C) and Pa is the vapour pressure in the cold air, which is usually very low. The skin wettedness, w, ranges from completely dry (w = 0 or taken as 0.06 for skin) to completely wet with maximum potential evaporation (w = 1.0). The value of w is somewhere between 0 and 1 for a moderately sweating skin. So an approximation to the evaporative component is WCI_E = 16.5 w $Kv^{0.5}$ × (5.65 − 0.65) or as a simple approximation = 100 w $Kv^{0.5}$. Putting it all together the overall WCI = WCI_{R+C} + WCI_E = $Kv^{0.5}$ (33 − tar − 100 w). For dry skin and no solar radiation tar = ta and w =0. The equivalent chilling temperature for the above simplified equation with calm air assumed to be 1.0 ms^{-1}, ta = tr and dry skin so w = 0, is tch = 33 − WCI/K.

The concept and experience of wind chill are mainly concerned with dry skin on body parts at low air temperatures in windy conditions. The term 'chill' is also used for cooling of wet or damp skin and for the whole body. For the purposes of the assessment of human cold stress and consequent cold strain, wind chill is confined to parts of the body. For whole-body assessment, all relevant variables are considered (air temperature, radiant temperature, air velocity, humidity, clothing and activity) and a full analysis using the heat balance equation, thermoregulatory response and criteria for physiological strain are used. This is the basis for the calculation of the IREQ index (Holmer, 1984; ISO 11079, 2007) discussed in the next chapter.

Shitzer and Tikuisis (2012) provide an analysis of existing wind chill and chilling temperature indices and conclude that whole-body models should be used. They also consider computer models of the face and fingers. They relate this to the application of the UTCI thermal index for weather conditions, with further research needed. It is a question of semantics and definition; however, it seems sensible to restrict wind chill to local discomfort and possible skin damage. Some whole-body computer models include body parts and could be used for both local and whole-body reactions. Shitzer and Tikuisis (2012). This is considered further in Chapter 9.

Required Clothing Insulation

8

CLOTHING FOR HEAT BALANCE

To maintain comfort and survival in cold conditions, people require clothing; not too much, not too little; but just the right amount!

Clothing is an integrated system (often mistakenly regarded as many independent items) and any clothing ensemble is designed to match environment, context and tasks as a component of a larger system. In cold conditions, the thermal insulation required of a clothing ensemble, in any environment, can be determined at the outset of design, from a calculation involving the human heat balance equation.

The human disposition as a homeotherm is to maintain an internal body temperature at around a constant value of 37 °C. A constant temperature means that there is no net heat gain or loss and hence heat balance. Heat can be transferred by conduction, convection, radiation and evaporation. Metabolic heat is produced by the body and clothing regulates the heat exchange between the body and the environment. For the clothing insulation to be 'just right' it will maintain body temperature and comfort by the avoidance of the need to sweat and by maintaining the mean skin temperature at around 33 °C (depending upon activity level (Fanger, 1970)). The heat balance equation and mechanisms of heat loss and gain are shown in Figure 2.1.

Gagge et al (1941) defined the clo unit as the clothing insulation required to maintain comfort for a resting person in still air at a temperature of 21 °C (1.0 clo = 0.155 m^2 °C W^{-1}). A nude person has zero clo and winter clothing will be around 2.0 clo. The military uniform was classified by clo, in a way that could easily be understood by military personnel and the heat balance equation was used to calculate the required clo values for cold conditions. For example a running soldier (6 mets) in air temperature of –34 °C requires a 1.5 clo uniform or a marching soldier (3.0 met) would require an air temperature of 4.0 °C if wearing 1.5 clo.

Burton and Edholm (1955) in their classic book *Man in a Cold Environment* criticize the wind chill index of Siple and Passel (1945) as 'it is impossible to express the effects of wind on heat loss without referring to the amount of clothing'. They propose an equivalent still air temperature (ESAT) where the insulation of the air around the body decreases in an exponential way as wind increases. That is from 1.0 clo in still air to 0.4 clo for wind at around 1.0 ms^{-1} to 0.1 clo at around 20 ms^{-1}. The equivalent temperature decrease is then estimated as a decrease of 0 °C (per met) for still air to 5.5 °C per met at around 1.0 ms^{-1} and 8 °C per met at around 20 ms^{-1}. The met is used here as a unit of heat loss (not directly related to the metabolic rate) where 1.0 met = 58.15 Wm^{-2}. A further refinement for radiation (cloud cover) led to the equivalent shade temperature (EST) and combining both (ESAT and EST) to the still shade temperature (SST). Further details are provided in Burton and Edholm (1955) and Parsons (2014).

Gagge et al (1941) and Burton and Edholm (1955) consider the required clothing insulation for the whole-body. Wind chill is considered as an effect on parts of the body where there is exposed skin, for example on the face, head and hands. A strategy for selecting appropriate clothing for activity in cold environments is to first calculate the clothing insulation required for the whole-body and then to provide additional clothing (mitts, gloves, hat, etc.) for the protection of the extremities.

Fanger (1970) integrated a human heat balance equation with criteria for thermal comfort into a comfort equation from which required clothing insulation for comfort can be calculated. This is valid for conditions near to comfort; however, for the assessment of cold stress we should turn to the Swedish scientist Ingvar Holmèr and his extensive work on the clothing insulation required index (IREQ).

THE IREQ INDEX

A simple starting point for estimating required clothing insulation is to use the temperature gradient between the body and the environment divided by metabolic rate. To maintain survival in the cold, heat transfer through clothing must balance with metabolic heat and heat transfer between the body and the environment. A simple representation would be metabolic heat production (H) matching heat conduction through clothing. That is H = K (tsk – tcl), where K is the thermal conductivity of clothing, tsk is the mean skin temperature and tcl is the mean temperature of clothing. The intrinsic clothing insulation Icl = 1/K; mean skin temperature can be estimated for comfort as tsk = 33 °C and tcl is somewhere between tsk and ta, depending upon air velocity. So in wind (>5ms⁻¹) with light activity (100 Wm⁻²) and ta = tr = −10 °C (so estimate tcl to be around −5 °C); Iclreq = (33 − − 5)/100 = 0.38 m² °C W⁻¹ = 0.38/0.155 = 2.45 clo. If we double the metabolic rate we halve the required clothing level and so on.

This may provide a good 'rule of thumb' starting point but more detail is usually required. The estimate of tcl may be sufficient but a more detailed consideration will require an iterative procedure as the mean temperature of clothing will depend upon the insulation of clothing, which is what we are trying to calculate. It will also depend upon environmental radiation and heat loss through other mechanisms such as breathing, clothing pumping effects and evaporation from the skin as well as any insulative air layers around the body at low air speeds. Using tsk = 33 °C, an estimate of metabolic rate for the activity and an estimate (educated guess) of tcl (based upon ta, tr and wind) can provide a useful starting point for clothing design. A computer program and detailed analysis, however, will provide a more accurate determination.

ISO 11079 (2007) "ERGONOMICS OF THE THERMAL ENVIRONMENT – DETERMINATION AND INTERPRETATION OF COLD STRESS WHEN USING REQUIRED CLOTHING INSULATION AND LOCAL COOLING EFFECTS"

Holmèr (1984) developed the method for determining IREQ and it is formerly represented in ISO 11079 (2007). The international standard specifies methods and strategies for assessing human cold stress due to indoor and outdoor cold environments. Cold stress is defined as "climatic conditions under which the body heat exchange is just equal to or too large for heat balance at the expense of significant and sometimes uncompensable physiological strain (heat debt)." Wind chill is related to the cooling effect on a local skin segment and IREQ is defined as "required clothing insulation for the preservation of body heat balance at defined levels of physiological strain".

The IREQ index can be used as a general cold stress index that indicates the effects of a cold environment on people as well as a method for calculating and selecting required clothing insulation and for calculating safe exposure times if a person wears less than the required level of clothing insulation. For local cooling the standard also presents a wind chill temperature for assessing convective cooling (see Chapter 7) as well as conductive cooling, extremity cooling and airway cooling.

CALCULATION OF THE IREQ INDEX

IREQ is calculated from the heat balance equation M-W = Eres + Cres + E + K + R + C + S, where all terms are in units of Wm^{-2}. M is the metabolic rate usually obtained from tables (ISO 8996, 2004) and W is external work, usually assumed to be 0. Eres is the loss of heat by breathing due to inspired cold air being heated and saturated in the lungs and the net loss of vapour in expired air. Cres is the loss of convective heat by breathing cold air where the air is heated to lung temperature then expired. E is evaporative heat loss at the skin and K is assumed to be insignificant for the whole-body heat balance equation (although highly significant for local skin contact with cold surfaces – see chapter 14) and so is assumed to be 0. R is net heat loss by radiation and C is net heat loss by convection. Positive heat transfer values are taken as heat loss and negative values heat gain (e.g. from the sun). For heat balance there is no net heat storage (S = 0).

The clothing required for heat balance (IREQ) is determined by adding up net losses to balance metabolic heat and heat loss to the environment. So M-W-Eres – Cres – E = (tsk – tcl)/Icl = R + C. Icl required then becomes IREQ = (tsk – tcl) / (M—W – Eres – Cres – E). The mean skin temperature to determine the minimum clothing required (IREQmin - high cold strain) is 33.34 – 0.0354 M °C and for low strain and a neutral sensation (IREQneutral) it is derived from tsk = 35.7 – 0.0285 M. For high cold strain the skin is assumed to be 'dry' (w = 0.06) but for comfort and neutral sensation it is related to the metabolic rate as w = 0.001 M, as it is considered an active person is comfortable with some sweating (Fanger, 1970). It is important to note that in fact the calculation provides required resultant clothing insulation (Icl,r). Icl does not take account of practical factors such as heat loss due to movement and air exchange due to pumping of clothing through vents. For that reason the required Icl value is usually greater than the IREQ value and a 'correction' is made.

COMPUTER PROGRAM FOR THE CALCULATION OF IREQ

A computer program for the calculation of IREQ, local cooling and safe exposure times is provided in an annex to an early version of the standard in the program language BASIC, and in the current standard with reference to a web site to run a program written in Java script (http://www.eat.lth.se/fileadmin/eat/ Termisk_miljoe/IREQ2009ver4_2.html). A description of how to construct a program in similar code (VBA) is provided step by step below for use with the, widely and internationally available, spreadsheet Excel. The programming environment is downloaded on request, within Excel, as an option on the spreadsheet menu. [*File; options; customize ribbon; tick Developer box; OK* - worked for me]. Use of *Macro* then allows easy transition between writing and editing the program and running the program on the spreadsheet. I used Microsoft Excel 2010, internet tutorials and referred to the book by Jelen and Syrstad (2010). Although different computer languages are available they often have similar structures (functionality) and the following code has general application.

Five practical questions related to the assessment of cold stress are answered by ISO 11079 (2007). These are:

1. What is the minimum clothing insulation needed to survive (but be under high cold strain)?
2. What is the clothing insulation required for comfort (or low cold strain)?
3. For a given level of clothing, what is the maximum exposure time for survival?
4. For a given level of clothing, what is the maximum exposure time for comfort?
5. What is the effect of wind chill and how will it limit safe exposure times?

Required clothing insulation values are IREQ; exposure times limited by criteria are Duration Limited Exposures (DLE) and wind chill temperature is given the symbol twc.

THE COMPUTER PROGRAM INTERFACE

The interface for the software provides the inputs and displays the outputs. An example is provided in Figure 8.1. This can be generated in the computer program or can be produced using normal Excel functions as a framework within which the computer program can operate.

	A	B		C	D	E	F		G	H	I
1	ISO STANDARDS for the assessment of Cold stress										
2											
4	ISO 11079: Calculation of the clothing required for comfort and survival.										
7											
8	ta	tr		v	rh	M	IREQneutral		Icl neutral	IREQmin	Icl min
9		-23		-23	0.5	20	150		2.60	2.90	2.28	2.53
10	Calculation of exposure times (mins) for comfort and survival, when wearing available clothing										
12	ta	tr		v	rh	M	Available clothing		DLEneutral		DLEmin
13		-23		-23	0.5	20	150		2.5	120		480
14	ISO 11079 - Prediction of Local cooling of parts of the body due to wind chill										
16	Air temperatu Air velocity (10m up) (kmh-1)							Wind chill temp			
17	ta	V10						twc			
18	-23.0	5.00						-28			
19	ISO 11079 - Interpretation										
22	Required clothing insulation to maintain comfort for 8 hour shift								2.9 clo		Calculate
23	Minimum required clothing insulation for conditions for 8 hour shift								2.5 clo		
24	Maximum exposure times for comfort when wearing clothing available								120 mins		
25	Maximum exposure times to ensure survival								480 mins		
26	Sensation caused by wind chill to local parts of the body						Very cold, risk of skin freezing				

FIGURE 8.1   Spreadsheet interface for computer programs related to ISO 11079 (2007).

COMPUTER PROGRAM TO DETERMINE IREQ_{MIN}

"IREQmin defines a minimum thermal insulation required to maintain body thermal equilibrium at a subnormal level of mean body temperature. The minimum IREQ represents some body cooling, in particular of peripheral parts of the body. With prolonged exposure extremity cooling may become a limiting factor for duration of exposure." ISO 11079 (2007).

Cold strain criteria for IREQmin are mean skin temperature $tsk = 33.34 - 0.0354$ M °C and skin wettedness $w = 0.06$.

INPUTS

It can be seen from the spreadsheet in Figure 8.1 that inputs to calculate IREQmin are air temperature (ta °C); mean radiant temperature (tr °C); air velocity (v ms^{-1}); relative humidity (%) and metabolic rate (M Wm^{-2}). The values shown can be replaced by the values of interest on the spreadsheet and taken as inputs using the following code (in VBA code ' : ' indicates a new line).

Range("A9"). Select: ta = ActiveCell.value: ActiveCell.value = ta

The example given is for air temperature, the computer code symbol is ta and the value is taken from the spreadsheet. ta then takes that value (e.g. $-23\ °C$ in Figure 8.1). Values for tr; v; rh; and M are then taken from the spreadsheet using similar code. Available intrinsic clothing insulation Icl = 2.5 is used to calculate duration limited exposure times and not to calculate required clothing.

For this program assume mechanical work WK = 0; clothing pumping rate $p = 8\ ls^{-1}$ and not walking so walksp = 0.

> If M <= 58 Then M = 58 : If M > 290 Then M = 290: If ta >= 10 Then
> ta = 10

Sets M at a minimum resting value of 58 Wm^{-2} and maximum value at an active 290 Wm^{-2} with a maximum air temperature of 10 °C.

> If walksp <= 0.0052 * (M − 58) Then walksp = 0.0052 * (M - 58) : If
> walksp >= 1.2 Then walksp = 1.2

Sets a minimum, 0.0052 * (M − 58), and maximum, 1.2 level for the variable walksp.

> If v < 0.4 Then v = 0.4 : If v >= 18 Then v = 18

Sets air velocity at a minimum of 0.4 ms^{-1} and maximum of 18 ms^{-1}.

CALCULATIONS

> tsk = 33.34 − 0.0354*M : wetness = 0.06 : tex = 29 + 0.2*ta

Sets mean skin temperature (tsk) for IREQmin (high strain); skin wetedness for no sweating and estimates temperature of expired air (tex).

> pex = 0.611 * Exp((17.27 * tex) / (237.3 + tex))
> psks = 0.611 * Exp((17.27 * tsk) / (237.3 + tsk))
> If (ta > 0) Or (ta = 0) Then
> Pa = (rh / 100) * 0.611 * Exp((17.27 * ta) / (237.3 + ta))
> Else
> Pa = (rh / 100) * 0.611 * Exp((21.875 * ta) / (265.5 + ta))
> End If
> Psa = (100 / rh) * Pa

The saturated vapour pressure at the skin and expired air and air temperatures are calculated along with the partial vapour pressure in the air which is related to relative humidity (Pa = (rh/100) × Psa).

$$Icl = Icl * 0.155 : Iar = 0.092*Exp(-0.15*v-0.22*walksp) - 0.0045$$

Sets basic insulation (Icl) from clo units to $m^2 \, °C \, W^{-1}$ and adjusts the insulation of the air to take account of air velocity and walking speed to give resultant insulation of the air (Iar).

$$IREQ = 1 : factor = 1 : ArAdu = 0.77$$

Sets starting values and radiation surface area for a standing person (0.77) to use stepwise iteration to determine IREQmin ($m^2 \, °C \, W^{-1}$); Rt ($m^2 \, kPa \, W^{-1}$); fcl (nd); and hr ($W \, m^{-2} \, °C^{-1}$).

$$350 \; fcl = 1 + 1.97*IREQ$$

Starts iteration at label 350 and calculates first fcl (ratio of clothing to nude surface area).

$$Rt = (0.06/0.38)*(Iar+IREQ) : E = wetness*(Psks - Pa)/Rt$$

Evaporative resistance of clothing (Rt) is wettedness (dry skin) divided by im for normal clothing (0.38) and multiplied by the total clothing resistance (air + clothing insulation). Evaporative heat exchange is wettedness multiplied by the partial vapour pressure gradient between skin and air divided by the evaporative resistance of clothing.

$$Hres = 0.0173*M*(Pex - Pa) + 0.0014*M*(tex - ta)$$

Heat transfer through breathing (Evaporative + Convective)

$$tcl = tsk - IREQ *(M-W-E-Hres)$$
tcl is proposed as part of the iteration
$$hr = 0.0000000567*0.95*ArAdu*(((tcl+273)\char94 4)-((tr+273)\char94 4))/(tcl - tr)$$
hr is calculated using the tcl value proposed as part of the iteration
$$hc = 1/Iar - hr : R = fcl*hr*(tcl-tr) : C = fcl*hc*(tcl - ta)$$

Heat transfer by radiation and convection are calculated

$$Balance = M - WK - E - Hres - R - C$$

For tcl, hr and hence IREQ to be correct Balance should add to zero.

If (Balance>0) Then IREQ = IREQ – factor : If (Balance<0) Then
    IREQ = IREQ + factor
factor = factor/2

To achieve balance = 0, IREQ must be increased or reduced accordingly and we reduce factor to converge to the true value.

If Abs(Balance) > 0.01 Then GoTo 350

If balance is achieved within 0.01 then assume true values have been found and stop iteration (go on to next line of code). If not, then go back to label 350 with new values calculated in the previous iteration.

IREQ = ((tsk – tcl)/(R+C))/0.155 : IREQM = IREQ

IREQ is calculated using the value for tcl from the iteration. This is the minimum (high strain) required clothing insulation (IREQM). We now need to work out what intrinsic clothing insulation is required as we have calculated a resultant insulation that includes practical effects such as pumping caused by movement and walking.

We split a complicated equation into parts (dummy variables I1 and I2) to calculate the minimum intrinsic (basic) clothing insulation required. Note that the ISO standard contains a mistake so the constant 0.15 in the standard is changed to −0.15 in the computer program.

I1 = IREQM + (0.092*Exp(−0.15*v−0.22*walksp) − 0.0045)/fcl
I2 = I1/(0.54*Exp(0.075*Log(p) − 0.15 * v − 0.22 * walksp) − 0.06 *
    Log(p) + 0.5)
ICLM = I2 − 0.085/fcl

OUTPUTS

In program development it is useful to display and check all variables including those that were inputs to ensure no errors occurred. For the final interface the values required should be displayed.

Range("H9").Select: ActiveCell.Value = IREQM: Range("I9").Select:
    ActiveCell.Value = ICLM

Places minimum resultant clothing insulation required (IREQM) in cell H9 and the clothing that would have to be selected from tables (intrinsic clothing insulation, ICLM) and hence would provide the required (resultant) clothing insulation in cell I9.

The code for the calculation of IREQM and ICLM provided above provides the basis for the calculation of additional values (IREQneutral; DLEmin; DLEnetral). Additions and modifications are provided below to answer specific questions. The additions and modifications can be added to the code above to provide one computer program or separate computer programs, depending upon preference.

DETERMINATION OF IREQneutral

"IREQneutral is defined as the thermal insulation required to provide conditions for thermal neutrality i.e thermal equilibrium maintained at a normal level of mean body temperature. This level represents none or minimal cooling of the human body." ISO 11079 (2007).

Cold strain criteria for IREQneutral are mean skin temperature tsk = 35.7 − 0.0285 M °C and skin wettedness w = 0.001 M.

The computer program code is therefore identical to that for the determination for IREQmin but the lines

tsk = 33.34 − 0.0354*M : wetness = 0.06 : tex = 29 + 0.2*ta are replaced
with
tsk = 35.7 − 0.0285*M : wetness = 0.001*M : tex = 29 + 0.2*ta

That is skin temperature and wetness criteria for comfort (neutral sensation). Not the high strain but survival (at the expense of uncomfortable skin temperature and cool to cold sensations), provided for IREQmin. Outputs for IREQneutral and Iclneutral replace IREQmin and Iclmin outputs/labels. For the interface in Figure 8.1, both options have been included in the computer program.

DETERMINATION OF DLEneutral

DLEneutral provides the maximum exposure time for comfort.

If the clothing required is worn then it is expected that people will be able to sustain an exposure to cold under high strain (IREQmin) or low strain (IREQneutral). If the required clothing is not available then the exposure time must be limited to acceptable levels, depending upon the clothing insulation available. This is calculated from a prediction of when the body will lose heat (negative heat storage) at a level beyond which becomes unacceptable. For ISO 11079 (2007) this is at a value of Q_{lim} = 144 kJ m^{-2}. This can be determined from the calculations provided above. Heat storage is calculated from the heat balance equation S = M – W – Eres – Cres – E – R – C. Dlim is then Q_{lim}/S and we replace the code tcl = tsk – IREQ *(M-W-E-Hres) with tcl = tsk – Iclr *(M-W-E-Hres - S) in the iteration loop where Iclr is the actual clothing insulation worn in the actual conditions.

For the conditions in Figure 8.1 it can be seen that the IREQmin is 2.5 clo (DLEmin = 480 mins = 8 hours), IREQneutral is 2.9clo and if only 2.5 clo is available then the exposure time must be limited to DLEneutral = 120 minutes.

DETERMINATION OF DLEmin

The default value for Dlim is based upon the IREQneutral condition; however, other criteria may be used, for example DLE min. The additions and adjustments to the computer code given for IREQmin are then similar to the adjustments given in the previous section. IREQmin in Figure 8.1 is given as 2.5 clo, sufficient clothing is now available and DLEmin = 480 mins or no restriction on an 8-hour shift.

DETERMINATION OF t_{WC}

t_{WC} = 13.12 + 0.6215ta – 11.37$V_{10}^{0.6}$ + 0.3965ta $\times$ $V_{10}^{0.16}$ where t_{WC} is the wind chill temperature (ISO use the symbol 't_{WC}') in °C; ta is air temperature (°C) and V_{10} is the 'meteorological' air velocity (kmhr^{-1}) measured on top of a 10m pole and is approximated by V_{10} = v $\times$ 1.5 for air velocity (v) at 1.5m height (Note that 10 kmhr^{-1} = 6.25 mph = 2.8 ms^{-1}).

The VBA code is

t_{WC} = 13.12 + 0.6215*ta – 11.37*V_{10}^0.16 + 0.3965*ta*V_{10}^0.16

ISO 11079 (2007) defines the Wind Chill Temperature (t_{WC}) as "the ambient temperature, which at a wind speed of 4.2 kmhr^{-1} produces the same cooling power (sensation) as the actual environmental conditions". The standard provides four categories of risk: (1) Uncomfortably cold (t_{WC} between –10 to –24 °C); (2) Very cold. Risk of skin freezing (t_{WC} between

−25 to −34 °C); 3. Bitterly cold. Exposed skin may freeze in 10 minutes (t_{WC} between −35 to −59 °C); 4. Extremely cold. Exposed skin may freeze within 2 minutes. (t_{WC} = −60 °C and colder). Figure 8.1 calculates t_{WC} to be −28 °C for ta = −23 °C and V_{10} = 0.5 ms^{-1} providing an interpretation of very cold and a risk of skin freezing.

The limit for high strain is t_{WC} = −30 °C and for low strain of t_{WC} = −15 °C. Finger temperature limits are 15 °C and 24 °C for high and low strain criteria respectively and for breathing for low activity (M<= 115 Wm^{-2}) air temperatures of −40 °C and −20 °C for high and low risk, respectively, and for high activity (M>115 Wm^{-2}) −30 °C and −15 °C for high and low risk, respectively.

ISO 11079 (2007) provides a pragmatic method for assessment of cold stress based upon the calculation of clothing required, safe exposure times, chilling of body parts and recovery times. ISO 12894 (2001) provides guidance on medical screening of workers who are to be exposed to cold ISO 15743 (2008) a risk assessment strategy for consideration of work in the cold and DIN 33403-5 (1994) for Ergonomics design.

The following chapter describes how a computer model can include not only whole-body responses of a clothed person to cold but also the inclusion of body parts (e.g. hands and feet), an important development when predicting human responses to cold environments.

A Computer Model of Human Response to Cold including Torso, Limbs, Head, Hands and Feet

9

COMPUTER MODELS OF HUMAN THERMOREGULATION

The development of the computer led to dynamic and integrated representations of the human body, human thermoregulation and heat transfer between the body and the environment. Some of the computer models represented the body to include body parts such as the head, hands, feet and more and therefore have application in cold environments. A prediction of body core temperature as well as hand skin temperature, for example, will have utility in assessing how cold stress can affect the health, comfort and performance of people.

Eugene Wissler from the University of Texas, USA, was one of the first modern pioneers of whole-body human thermal (computer) modelling (Wissler, 1961) and in a review (Wissler, 1984) cites Whyndam and Atkins (1966) and Stowijk and Hardy (1966) who proposed models. Some early versions of models were represented on analogue computers (computers that used electrical circuits rather than digital techniques).

The first models for the digital computer were developed when Wissler (1961) for the US Air Force and Stolwijk (1971) for NASA, produced code. FORTRAN was the engineering computer language at the time and code was represented on computer cards (perforated tape, etc.), one card per line of code so prestige was carrying a large box of cards to the computer centre and returning days later to collect reams of computer paper containing output. The processing and administrative time often took over 24 hours to run on the most powerful computers at the time so this was a large organisation activity.

As the digital computer developed into the personal computer, models became more accessible. Parsons and Haslam (1984) and Haslam and Parsons (1989a, 1989b) were the first to take a pragmatic approach to application. They recognised the generic nature of the models and resisted producing yet another model but instead modified prominent existing models and evaluated them by comparing their predictions (skin temperatures, sweat loss, etc.) with the responses of human subjects. Models that were evaluated included a model of Vogt et al (1981) based upon a heat balance equation used for the assessment of heat stress (Belding and Hatch, 1955) and developed in Strasbourg, France (later to be incorporated into ISO 7933 (1989, 2004); two models from the J B Pierce Foundation Laboratory, Yale University, New Haven, Connecticut, USA; the 25-Node model presented by Stolwijk and Hardy (1977) and the 2-node model of Nishi and Gagge (1977); and the model of Givoni and Goldman (1972) developed empirically for the US Army. The work of Gordon (1974) and Ringuest (1981) was also considered.

Much of the research of Haslam and Parsons (1989a) was inspired by Dr Mike Haisman for the Army Personnel Research Establishment in the UK (along with numerous Loughborough students in and out of the thermal chamber) and led to a model system, the proposal for a toolbox to allow model building for any application and the use of finite difference and finite element methods using engineering software (Haslam and Parsons, 1989a; Wadsworth and Parsons, 1989; Neale, 1998). Later work to the present day includes the models of Fiala (1998) and Fiala et al (2012) as well as Brode et al (2012) and the large database approach of Parsons and Bishop (1991).

Parsons (2014) notes that any model will be an approximation to reality, "whether a model 'works' or not becomes a question of whether the imperfections are significant in terms of the application to which the model is put". This is the spirit in which models are presented here. A landmark model that forms the basis of most human thermal computer models and written in FORTRAN was that of Stolwijk and Hardy (1977) based on work for NASA in the 1960s. This has been adopted to produce many other models for example that of Gordon (1974), for the assessment of cold stress, and Fiala (1998) and Fiala et al (2012). A description and modified version of the Stolwijk and Hardy model is provided below and written in the internationally available VBA code that can be used in the spreadsheet of Excel 2010.

MULTI-NODE MODELS FOR THE ASSESSMENT OF COLD STRESS

The 'state' of an object can be represented by a collection of interrelated points that each have values (e.g. temperatures) and as time moves forward those values can change with relevant variables in a way determined by the laws of physics, including those of thermodynamics. A human thermal model is represented by such points or nodes and each point represents the state of part of the body. A mesh or network of closely knit points leads to finite element, finite difference or finite volume methods and is solved by a matrix of equations. If the point or node is representative of the 'state' of a whole region of the body lumped together, the models are often called 'lumped parameter models'.

Stolwijk and Hardy (1977) represent the body in 25 nodes or compartments representative of 6 segments (head; trunk; arms; hands; legs; and feet) × 4 layers (compartments-core; muscle; fat; skin) and a blood compartment that perfuses all other compartments. That is the $6 \times 4+1 = 25$ Node model. The dynamic behaviour of the model is determined by the structure and properties of a passive system, representing the morphology, anatomy and physiology of the body (the controlled system), a system of thermoregulation (the controlling system) and the heat transfer between the nodes (compartments in the body) and the environment.

The passive system is represented by concentric cylinders for the trunk, arms, hands, legs and feet, and the head by concentric spheres (see Figure 9.1). The model is symmetrical, with one cylinder with doubled values representing each pair of arms, hands, legs and feet.

The system is defined by physical dimensions as well as the thermal properties of tissues and blood flow. The controlling system involves sensors in the skin that feed information concerning skin temperatures to an integrated central processing site that 'decides' on an appropriate output that influences physiological responses simulating vasodilation, vasoconstriction, sweating and shivering and hence skin temperatures and wetness across the body as well as core temperature. Heat transfer is determined within the body between compartments including blood, and with the environment by convection, radiation and evaporation. The dynamic nature of the model is simulated by starting at a set of initial conditions (e.g. for comfort) and adjusting those conditions over time (e.g. every minute) depending upon the consequences of the physiological responses. Vasoconstriction will reduce skin temperature and hence heat loss to the environment that may stem a drop in core temperature. If it does not, then shivering will start by increasing metabolic heat production and so on.

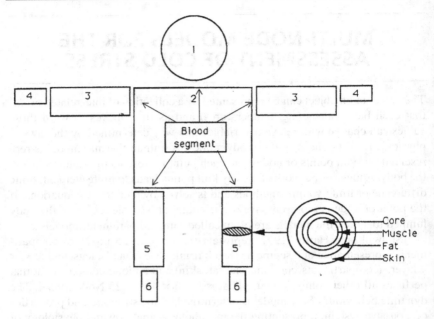

FIGURE 9.1   Diagrammatic representation of Stowijk and Hardy 25-Node model of human thermoregulation. From Haslam and Parsons (1989a).

A 'STOWIJK AND HARDY' MULTI-NODE COMPUTER MODEL OF HUMAN RESPONSE TO THE COLD

Stolwijk and Hardy (1977) provide FORTRAN computer code for a 25-Node model of human thermoregulation for a nude person. Haslam and Parsons (1989a) provided FORTRAN code for a clothed version. Galway and O'Sullivan (2005) provide a review of use of the models in computer-aided design. Procedural computer languages such as FORTRAN and BASIC are well suited to the versions of the model presented. An object-orientated approach (e.g. using VBA (Visual Basic Application) and an Excel spreadsheet) allows a flexible interface, almost universally available and with a platform of continuously updated facilities.

Stolwijk and Hardy (1977) present a flow diagram of their model in 12 stages that they use to structure their computer code. These are: read constants for controlled systems; read controller constants; read initial conditions; read experimental conditions; establish thermo-receptor output; integrate peripheral afferents; determine efferent outflow; assign effector output; calculate heat flows; determine integration step; calculate new temperatures: and if the required time has elapsed, present the output. The computer code is presented in the following paragraphs and in Appendix 1 in VBA macro code and Excel 10 – see Chapter 8).*

INTERFACE

A version of a computer interface is presented in Figure 9.2. This shows all inputs required and relevant outputs across the body as they change over time for a cold exposure. As it is a spreadsheet format these can be changed to match the characteristics of the simulated person exposed (tall, short, fat, thin, muscular, male, female, child, baby, disabled, ill, dead and more) as well as modified for different environments (indoors, outdoors, in and under water, up mountains, in space, on planets and more).

CONSTANTS FOR CONTROLLED, CONTROLLER AND ENVIRONMENTAL SYSTEMS

Compartments or nodes are referred to as N (a total of $6 \times 4 + 1 = 25$) and segments as I (= 6). Layers are given the values K= 3 for core; 2 for muscle; 1 for fat and 0 for skin. So segments are numbered, $4 \times I - K$, hence trunk core is N = 5 ($4 \times 2 - 3$) and so on. Blood compartment N = 25. It is necessary at the outset to define variables using the Dim statement as in Dim C(25). This defines the internal computer memory required for C. Similar statements are used for Dim T(25) the temperature of the 25 compartments: Dim TSET(25) the set point temperatures: Dim RATE(25) the sensitivity to rate of change of temperatures: Dim C(25) heat capacitances: Dim QB(24) basal metabolic heat productions: Dim EB(24) basal evaporative heat losses: Dim BFB(24)

---

* To set up the developer to develop and run macros in Excel see chapter 8. [*File; options; customize ribbon; tick Developer box; OK* - worked for me]

HEAD	1 CORE	2.57	14.95	0	45	1.61	36.96	0	36.96			HEAD	TRUNK	ARMS	HANDS	LEGS	FEET	CONTROL		
	2 MUSCLE	0.39	0.12	0	0.12	13.25	35.07	0	35.07		S	0.13	0.68	0.25	0.09	0.60	0.13	CSW	372.00	
	3 FAT	0.26	0.13	0	0.13	16.1	34.81	0	34.81		HR	4.80	4.80	4.20	3.60	4.20	4.00	SSW	33.70	
	4 SKIN	0.28	0.09	0.81	1.44	0	34.58	0	34.58		HCB	3.00	2.10	2.10	4.00	2.10	4.00	PSW	0.00	
TRUNK	5 CORE	11.44	52.63	10.45	210	1.59	36.89	0	36.89		ARAD	0.74	0.75	0.66	0.56	0.65	0.62	CDIL	136.00	
	6 MUSCLE	18.8	5.81	0	6	5.53	36.28	0	36.28		SKINR	0.21	0.42	0.10	0.04	0.20	0.03	SDIL	8.90	
	7 FAT	4.94	2.49	0	2.56	23.08	34.53	0	34.53		SKINS	0.08	10.48	10.15	40.03	10.21	80.04	PDIL	0.00	
	8 SKIN	1.41	0.47	3.78	2.1	0	33.62	0	33.62		SKINV	0.13	0.32	0.1	0.12	0.23	0.10	CCON	10.80	
ARMS	9 CORE	1.63	0.82	0	0.84	1.4	35.53	0	35.53		SKINC	0.01	0.05	0.19	0.20	0.20	0.35	SCON	10.80	
	10 MUSCLE	3.54	1.11	0	1.14	10.3	34.12	0	34.12		WORKM	0.00	0.30	0.08	0.01	0.60	0.01	PCON	0.00	
	11 FAT	0.67	0.21	0	0.2	30.5	33.59	0	33.59		CHILM	0.02	0.85	0.05	0.00	0.07	0.00	CCHIL	13.00	
	12 SKIN	0.5	0.15	1.4	0.5	0	33.25	0	33.25									SCHIL	0.40	
HANDS	13 CORE	0.16	0.09	0	0.1	6.4	35.41	0	35.41	INPUTS	TTIME hr	1.00						PCHIL	1	
	14 MUSCLE	0.07	0.23	0	0.24	11.2	35.38	0	35.38		ITIME min	TA	TR	V	RH	CLO	Iecl	WORK	WK	
	15 FAT	0.1	0.04	0	0.04	11.5	35.3	0	35.3		10	5	5	1	0.2	1	0.0155	100	0	
	16 SKIN	0.2	0.06	0.52	2	0	35.22	0	35.22	OUTPUTS	MTIME	Tcore	Tsk	T hand	T foot	w	MR	(C+R)	E	ESK
LEGS	17 CORE	4.94	2.59	0	2.69	10.5	35.81	0	35.81		0.00	36.96	33.97	35.22	35.04	0.06	100.00			
	18 MUSCLE	10.67	3.32	0	3.43	14.4	35.3	0	35.3		10.00	36.62	31.41	31.83	31.40	0.03	107.14	117.45	15.46	5.59
	19 FAT	1.66	0.5	0	0.52	74.5	35.31	0	35.31		20.00	36.58	30.49	30.35	29.50	0.03	111.58	113.21	15.46	5.59
	20 SKIN	1.25	0.37	3.32	2.85	0	34.1	0	34.1		30.00	36.57	29.96	29.53	28.26	0.03	113.94	110.83	15.46	5.59
FEET	21 CORE	0.27	0.15	0	0.16	16.3	35.14	0	35.14		40.00	36.58	29.66	29.09	27.50	0.03	115.43	109.35	15.46	5.59
	22 MUSCLE	0.07	0.02	0	0.02	20.6	35.03	0	35.03		50.00	36.57	29.44	28.83	26.97	0.03	116.44	108.39	15.46	5.59
	23 FAT	0.15	0.05	0	0.05	16.4	35.11	0	35.11		60.00	36.57	29.29	28.68	26.63	0.03	117.18	107.75	15.46	5.59
	24 SKIN	0.26	0.08	0.72	3	0	35.04	0	35.04											
	25 BLOOD	2.6					36.71	0	36.71											
Property:		C	QB	EB	BFB	TC	TSET	RATE	T											

FIGURE 9.2 Spreadsheet interface for the modified Stolwijk and Hardy 25-Node model of human thermoregulation.

basal effective blood flows: Dim TC(24) thermal conductance between N and N+1: Dim S(6) surface area of segments: Dim SKINR(6) fraction of all skin receptors in segment I: Dim SKINS(6) fraction of sweating command for skin of segment I : Dim SKINV(6) fraction of vasodilation command applicable to skin of segment I: Dim SKINC(6) fraction of vasoconstriction command applicable to skin of segment I: Dim WORKM(6) fraction of total work done by muscles in segment I: Dim CHILM(6) fraction of total shivering occurring in muscles in segment I : Dim HR(6) Radiant heat transfer coefficient for segment I: Dim HCB(6) basic convective heat transfer coefficient for segment I: Dim HC(6) convective heat transfer coefficient for segment I: Dim H(6) total heat transfer coefficient for segment I: Dim F(25) rate of change of temperature in N : Dim WARM(25) output from warm receptor N : Dim COLD(25) output from cold receptor N: Dim HF(25) rate of heat flow into or from N: Dim DRIVER(25) output from thermo-receptors in compartment N: Dim Q(24) total metabolic heat production in N: Dim E(24) total evaporative heat loss from N: Dim BF(24) total effective blood flow to N: Dim EMAX(6) maximum 'possible' rate of evaporative heat loss from segment I: Dim BC(24) convective heat transfer between central blood and N: Dim TD(24) conductive heat transfer rate from compartment N to N+1: Dim ARAD(6) radiant surface area for segment I : Dim DRY(6) total dry heat transfer for segment N: Dim RT(6) basic or intrinsic evaporative resistance of clothing for segment I : Dim TCL(6) mean surface temperature of clothing for segment I: Dim LR(6) Lewis relation for segment I.

The first line of the computer code for the VBA program 'Stolwijk&Hardy' is therefore

```
Dim T(25): Dim TSET(25): Dim RATE(25): Dim C(25): Dim QB(24):
    Dim EB(24): Dim BFB(24)
```

Followed by other Dim statements. These are then read from the interface with the following looped code. Note that I have separated code instructions with ' : '. For effective layout, separate lines indented between 'For' and 'Next' may be convenient.

```
For I = 1 To 25: Cells(I, 3).Select: C(I) = ActiveCell.Value: Next I
```

This will read the list of 25 C values from the 3rd column and 25 rows of the interface (Figure 9.2). All variables with 25 values are read in this way and loops with 24 and 6 values are read in a similar modified way.

Single values are read from individual cells for example for the control coefficient for sweating from head core, placed in cell row 6 column 11. Air temperature TA is in row 15, column 13 and so on.

Cells(6, 11).Select: CSW = ActiveCell.Value : Cells(15, 13).Select:
TA = ActiveCell.Value

CSW, sweating from head core; SSW, sweating from skin; CDIL, vasodilation from head core; SDIL, vasodilation from skin; CCON, vasoconstriction from head core; SCON, vasoconstriction from skin; CCHIL, shivering
from head core; SCHIL, shivering from skin; PSW, sweating from skin and
head core; PDIL, vasodilation from skin and head core; PCON, vasoconstriction from skin and head core; PCHIL, shivering from skin and head core;
BULL, factor determining temperature sensitivity of sweat gland response.
Environmental conditions are air temperature, TA; Mean radiant temperature, TR; air velocity, V; and relative humidity, RH. Clothing is input as a
CLO value for intrinsic insulation and an Iecl value for intrinsic evaporative
resistance (see ISO 9920, 2007).

UNITS

Units are in Watts for heat flow, °C for temperature and hours for time (heat
capacitance of compartment 1 (head core) is for example 2.57 W h °C^{-1} - see
cell C1 in Figure 9.2). These are converted to minutes and seconds as appropriate and divided by surface area to give Wm^{-2} or vice versa when required.
Partial vapour pressure and saturated vapour pressures are related to humidity
and estimated using Antoine's equation (see later and Chapter 8) to provide
Pa and kPa and determine heat flow by evaporation. Clo values are converted
using 1 Clo = 0.155 m^2 °C W^{-1}. Timers for the dynamics of the program are
in minutes, although other values are possible depending upon rate of change
of values.

The interface provides values and the procedures and code above provide mechanism for reading constants for the controlled (passive) system that
defines the body; controller constants that determine the responses of the
body to given conditions for example the drivers to enable the physiological
responses to restore any differences between the temperatures of the body
compartments and the 'optimum' set point temperatures; the starting conditions of the body for example body compartment temperatures; and the environmental conditions to which the model is exposed that determine the heat
transfer between the body and the environment. Some preliminary calculations are then made as follows before the thermoregulatory loop and dynamic
modelling can begin.

These calculate the total surface area of the body by adding up compartment surface areas

SA = 0 : For K = 1 To 6 : SA = SA + S(K): Next K

adding a small number to the CLO value to ensure there is no division by 0 for a nude person, a calculation of the increase in body surface area due to clothing (FACL), setting skin wettedness at a starting insensible perspiration value of 0.06 and using MR as the symbol for metabolic rate (initial input as WORK in Wm^{-2}).

If (CLO = 0) Then CLO = CLO + 0.0001: FACL = 1 + 0.31 * CLO: PWET = 0.06: MR = WORK

They include converting metabolic rate (WORK) and mechanical work output (WK) from Wm^{-2} to W and setting sensitivity to rate of change to 0.

WORK = WORK * SA : WK = WK * SA : For N = 1 To 25: F(N) = 0: Next N

Initial mean skin temperature is also calculated

TS = 0 : For I = 1 To 6: TS = TS + T(4 * I) * C(4 * I) / 3.9: Next I

If the muscles are not at rest they will produce additional heat related to mechanical work (WK) and metabolic rate (WORK) above that of resting (86.49 W).

WEFFM = 1 - (WK / (WORK - 86.49)): WORKI = (WORK - 86.49) * WEFFM: If (WORK <= 86.49) Then WORKI = 0

Finally, the convective heat transfer coefficient for the segments of the body can be calculated

For I = 1 To 6 : HC(I) = 3.16 * HCB(I) * V ^ 0.5 : Next I

The outputs of the model provide selected physiological conditions of the body over a time of interest, displaying the results at selected intervals (e.g. every 10 minutes for 1 hour). To begin the display the starting conditions are presented. Exposure time (MTIME), head core temperature (T(1); mean skin temperature (TS); hand skin temperature (T(16)); foot skin temperature, T(24); skin

Wettedness, PWET; metabolic rate, MR; and dry heat exchange, DRYT, are available and presented as starting conditions (MTIME = 0). For a given activity and clothing, heat flows into and out of the body depending upon environmental conditions. Body temperatures change and the new conditions feed into the model to generate, in a loop, the change in the model condition over time to a final exposure time.

The initial conditions are presented using code similar to the following for initial head core temperature (T(1) placed into the cell in row 19, column 13 (see Figure 9.2)

```
Cells(19, 13).Select: ActiveCell.Value = T(1)
```

All of the above are values that do not change with time. In the code, during development, I printed them onto sheet 2 to exactly match the inputs from sheet 1 and also outputs from other preliminary calculations. By default we start in sheet 1 and an example for the 25 values of C (heat capacitances) to be read then a copy placed in sheet 2 is:

```
For I = 1 To 25 : Cells(I, 3).Select : C(I) = ActiveCell.Value
Sheet2.Select : Cells(I, 3).Select : ActiveCell.Value = C(I) : Sheet1.
    Select : Next I
```

THE THERMOREGULATORY LOOP

The thermoregulatory loop calculates how the model of the body will change in terms of temperatures of the 25 compartments for example over an interval of time, taken in this version of the model as 1 minute (or an optimum calculated according to rate of change of values). The conditions after one interval are returned to the start of the loop and the calculations start again for the time period between one and two intervals, and so on until a final exposure time is reached. Initial conditions can be changed, for example a change in environmental conditions during the exposure by returning to the appropriate code in the program above. The code for the thermoregulatory loop is shown in Appendix 1.

OUTPUT OF PREDICTED HUMAN RESPONSES

Results selected as outputs from the program depend upon what is relevant to the application of interest. For human response to the cold, internal body temperature (brain head core, T(1)) will indicate overall heat loss, if any, and possible hypothermia. Mean skin (TS) and local hand (T(16)) and foot (T(24)) skin temperatures will indicate discomfort and local injury as well as possible effects on manual dexterity. Skin wettedness will indicate the state of the skin and metabolic heat will indicate shivering. Dry and evaporative heat flows will indicate the dynamic nature of the interaction. The code for presenting outputs is presented in Appendix 1 and outputs for an example of human response to cold stress (ta=tr=5 °C; v=1 ms^{-1} etc.) are shown in Figure 9.2.

Figure 9.2 shows the predicted outcomes based upon physiological responses over time from a starting thermal comfort condition. For 1 hour exposure with an output shown every ten minutes, the head core temperature decreases from 36.96 °C to 36.57 °C; mean skin temperature from 33.97 °C to 29.29 °C; hand skin temperature from 35.22 °C to 28.68 °C; and foot skin temperature from 35.04 °C to 26.63 °C. An interpretation could be that the person would survive but be uncomfortable with a probable loss in performance in tasks requiring manual dexterity. Improvement to conditions can be investigated with repeated runs of the model, investigating increasing clothing insulation, either the whole body or the use of gloves, increasing activity level and more.

Human Performance and Productivity in the Cold

10

HEALTH AND SAFETY, DISTRACTION AND CAPACITY (HSDC)

Being cold reduces human performance and can greatly affect productivity. Cold stress can reduce skin and muscle temperature and increase fluid viscosity (e.g. blood and synovial fluid), causing slowing of movements, reduced sensitivity and loss of strength. The discomfort and dissatisfaction caused by cold can cause distraction and 'time off task' and if there is a risk to safety (e.g. increased risk of accidents) as well as injury and hypothermia then work must cease and productivity can fall to zero. It should also be remembered that wearing clothes reduces performance and cold weather clothing is cumbersome to wear, reduces manual dexterity when wearing mittens and gloves and becomes laborious and even depressing when it has to be taken off and on or adjusted to accommodate bodily functions and activity.

There have been a number of comprehensive studies and reviews of the effects of cold on human performance. (e.g. Fox, 1967; Wyon, 2001; Parsons, 2014). Studies have mainly compared performance at standardised simple tasks in the cold with performance at the same task when comfortable or warm. There have also been practical studies and much experience gained in cold store exposures and expeditions to cold regions (Tochihara, 1998; Ashcroft, 2000). Parsons (2020) provides a methodology in the region of thermal discomfort where cold discomfort would probably not cause loss in productivity due to health and safety regulations but distraction caused by discomfort and loss in capacity at some tasks may. The early British Offices, Shops and Railway premises act (HMSO, 1963) specified that, after one hour of work, thermal environments must reach at least 16 °C air temperature for sedentary work and 13°C for active work. Cold has also been considered so uncomfortable and unacceptable that workers refuse to work.

Fox (1967) reviews studies of the effects of cold environments on human performance and identifies tactile sensitivity, manual performance, tracking, reaction time, complex behaviours, maintaining hand skin temperature and adaptation and acclimatisation as relevant. In summary it is concluded that hand skin temperature is a good predictor of loss of performance in the cold as cold mainly affects manual dexterity. McIntyre (1980) suggests that cold affects both manual performance (due to a loss of finger sensitivity and strength) and performance at complex mental tasks due to over-arousal. (Parsons (2020) would interpret this as causing hyperactivity for the task and distraction.) He notes that if muscle temperature is less than 27°C then strength will decrease with reduced grip strength to 40% of 'original' values when the hand is immersed in water at 2°C.

Sensitivity can be measured using pointed 'dividers' with two points placed unseen on bare skin (occasionally only one point is used to assess guessing). The ability to detect two points at different distances between them provides a measure. The closer the points (correctly identified as two), the more sensitive the skin, and the threshold is the largest distance below which, when two points are present, only one is felt. A modified version is the Mackworth V (Mackworth, 1953), where the finger is placed at random points over the V and the distance from the apex of the V, where the threshold occurs between perception of one edge and two is the measure. Provins and Morton (1960) use Mackworth's V to show that, for one subject, a critical temperature exists at 6°C finger skin temperature when numbness occurs but above which sensitivity is minimally affected. This critical temperature varies with individuals. The finer the dexterity required the greater the loss in manual dexterity. McIntyre (1980) suggests a lower critical temperature for loss in manual dexterity of between 13 and 16°C. Fox (1967) suggests 12°C16°C as a threshold for manual dexterity and 8°C for numbness. ISO 13732-3 (2005) uses 15°C as a threshold for manual dexterity and 5°C for numbness. Parsons (2014) cites Todd (1988) where it is suggested that hand skin temperatures below 30°C begin to cause a loss in performance at a block rotation task. This was part of a study to predict loss in performance from environmental conditions and predictions of the hand skin temperature from a modified version of the Stolwijk and Hardy (1977) 25 Node model of human thermoregulation.

Parsons (2014) considers the effects of cold on cognitive and manual performance. For cognitive performance, little direct effects are noted but it was found that distraction was prevalent. He cites Horvarth and Freedman (1947) and Enander (1987) who generally found that cold had little direct effect on capacity to perform cognitive tasks, in particular reaction time, but that distraction did occur and that Payne (1959) found subjects to show lack of attention, hostility towards the experimenter and a desire to withdraw from the experiment. For extreme conditions and conditions which elicit physiological change and even injury (or collapse), performance will be expected to fall 'catastrophically'. It is therefore important that health and safety regulations are observed. There have been reports of loss of memory for periods where people have been subject to hypothermia but these are exceptional and may be related to reduction in metabolic rate (see also Parsons, 2018).

Meese et al (1984) conducted experiments on factory workers in thermal chambers parked in factory car parks to ensure a link with the practical work and context. Hot, moderate and cold conditions were investigated. Cold sessions involved air temperatures of 6, 12, 18 and 24°C. Around 1,000 workers conducted 17 different tasks over 8-hour shifts. It was found that even for 'moderate' cold a fall in temperature from 24°C to 18°C caused a drop in manual dexterity and that performance at pegboard, screw plate, block threading, knot tying and assembly line tasks all decreased as temperatures fell from 24°C to 6°C.

ISO HUMAN PERFORMANCE FRAMEWORK

ISO TR 23454 (2020, 2021) parts 1 and 2 present a performance framework for estimating the effects of the physical environment on performance, including cold, and simple tasks and methods for quantifying those effects in terms of that framework (see also Parsons,2018).

The framework considers three mechanisms for describing the effects of the physical environment on human performance. (see Figure 10.1). These are the effects on *human function* or activity (reduced capacity to perform a task), *distraction* due to attending to the effect of the physical environment rather than the task or activity being performed and zero performance for a time, due to a *suspension of work* for health and safety reasons, where physical environments are beyond limits for safe working or other activity.

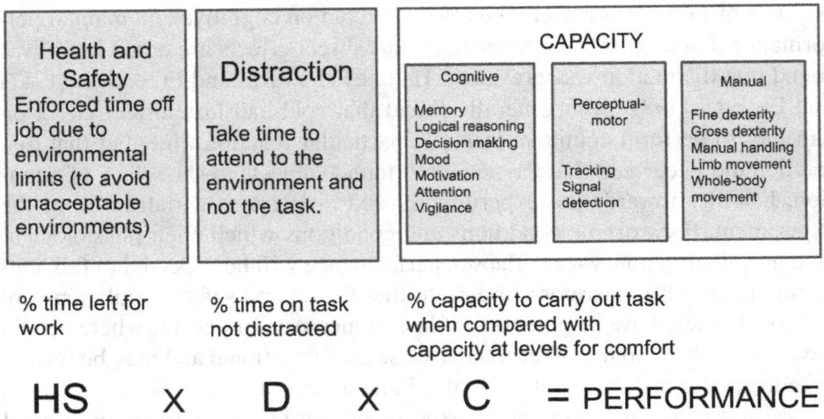

FIGURE 10.1   HSDC model for predicting the effects of the environment on human performance.

The effects of the physical environment on human function are divided into effects on cognitive, perceptual-motor and manual performance. Distraction is caused by a tendency to attend to the effect of the physical environment rather to the task or activity and hence will reduce performance because less time is spent on the task. When work is suspended, due to physical environments that will affect health and safety (or maybe because the environment is considered unacceptable for other reasons such as discomfort), then performance will reduce to zero.

EXAMPLE: HSDC AND PERFORMANCE IN THE COLD

An example of how the framework could be used in cold conditions is provided in an informative annex of the standard. It emphasises that it presents a mechanism, a method and realistic but not necessarily accurate estimates.

Workers are conducting a food preparation task in light clothing and a cool environment of 4 °C air temperature, 2°C mean radiant temperature, 0.5 ms^{-1} air velocity over the workers and 50% relative humidity. They are expected to carry out an 8-hour shift. Health and Safety limits are derived from international standards. ISO 11079 (2007) provides duration limited exposure times of 3.4 hours before the environment becomes unacceptable (DLEmin). Although it would be possible to return after recovery, workers do not generally find returning to a cold environment to be acceptable. Calculations can be made for total work in a work/rest cycle, but it will be assumed that health and safety considerations restrict work to 3.4/8 = 0.425 = 42.5%.

For the cold conditions, ISO 7730 will provide a Predicted Percentage of Dissatisfied of around 100% as conditions are beyond lower limits for comfort, and we can estimate that the workers will concentrate on the environment for 20% of the time as they are uncomfortably cold, so they are distracted for 20% of the time, and 80% (0.8) of the time is spent on work. As the food preparation involves mainly manual tasks with some dexterity we can estimate from studies with simple tasks in the cold that capacity will be reduced by 30% so that it is at 70% (0.7) of optimum levels.

Estimated performance for the cold conditions considered uses HS = 0.425; D = 0.8; C = 0.7 so HS × D × C = 0.425 × 0.8 × 0.7 = 0.238 or 24% of the performance that would be expected at optimum (comfortable) levels.

Increasing clothing level is an obvious solution to reducing the loss in performance (although wearing clothing can also reduce performance in terms of distraction and capacity). ISO 11079 (2007) provides a method for calculating the amount of clothing required for comfort and for the conditions in the above example an intrinsic insulation of 2.1Clo (see ISO 9920, 2007) provides a starting point for clothing design. Allowing workers to select their own clothing based upon that value is also beneficial. Providing appropriate clothing to workers will not only provide worker satisfaction but will increase productivity.

To achieve valid data on human performance for the particular cold conditions of interest it is useful to conduct user tests and trials with representative samples of human subjects. How to conduct those tests and trials will be described in ISO TR 23454 Part 2 (2021).

Diverse Cold Environments

11

COLD WATER

Most sea temperatures are less than 20 °C and most people in water have to combat cold. When considering human response to cold water a starting point is to consider human response to the environment as determined by metabolic heat gain being balanced by heat exchange by conduction, convection, radiation and evaporation. In water the heat transfer is mainly restricted to heat loss by conduction and convection. Naked people will cool up to 4x faster in cold water than in air at the same temperature. This is because the thermal conductivity of water is 24x that of air and its volume-specific heat capacity (specific heat × density) is 3,500x that of air (Golden and Tipton, 2002).

The critical temperature for a person can be defined as the temperature below which their resting metabolic rate starts to increase. The fatter the person, the higher the critical temperature. For air temperature this is between around 20°C and 27 °C and for water temperature between around 32 °C and 35 °C. A thermoneutral temperature in water is around 33 °C and below that there begins to be a net heat loss from the body to the water. Physiological responses include vasoconstriction and shivering but also Cold Induced VasoDilatation (CIVD) can occur which inconveniently promotes heat loss from the body when it is attempting to preserve heat.

Golden and Hervey (1981) identify four phases of response when a person is immersed in cold water. An initial response (0–3 min) with rapid respiration (a dangerous gasp), cardiovascular, cardiac and hormonal response; short term (3–30 min), where manual performance, strength, and ability to swim deteriorate; long term, where hypothermia can cause unconsciousness, drowning and cardiac arrest; and post immersion (and recovery), where care is required to avoid a rapid change in blood pressure, which may cause death.

For still water especially with appropriate clothing, a boundary layer of 'warmer' water can reduce heat loss. For moving water or when a person is active (swimming) heat loss is greatly increased (Keatinge 1969). For cool and cold water, for a person in a lifejacket, it is best to stay still (a pseudo-foetal position is recommended, Hayward et al, 1975).

SURVIVAL TIMES

Survival times in cold water have been estimated from thermal models (see chapter 9) as well as experience and empirical studies. Criteria for survival vary and have used a deep body temperature of 28 °C and estimations of times of useful consciousness (e.g. a drop to 35 °C in internal body temperature). Brooks (2001) suggests that it is the first two stages of initial immersion and swimming in cold water that account for many deaths and that survival times should, but tend not to, take those into account. Golden and Tipton (2002) provide a review (see also Oakley and Petherbridge, 1997). For lightly clad males they show 50% survival times or times of likely death ranging from 1 – 2.3 hours for 5 °C water temperature to 3–7.7 hours for 15 °C water temperature. They also note a range of confounding factors including clothing and degree of fatness. These factors could be taken into account in thermal models. Recommended search times are 3–6x those of estimated 50% survival times. For survival, a positive mental attitude is important. Immersion in water involves close to 100% of the body surface area involved in heat loss. Water movement due to currents and tides (tidal steams are available on charts), and additional factors related to rough seas all contribute to heat loss in cold water. An overriding interpretation of predicted survival times is that there are many individual and contextual factors that need to be taken into account and that the recommendations should be taken as a starting point modified (often significantly) by expert consideration of those factors.

UNDER THE WATER, IN THE WATER AND ON THE WATER

People cannot survive under water for long without diving equipment. In cold water, ability to hold breath is severely restricted as is the ability to swim. They are also subject to increased pressure by 1 additional atmosphere for every 10m of depth. Deep water temperature in the ocean remains at around a constant 2 °C and heat loss from the body is increased with decreasing temperature and increasing pressure. Even on the surface it is rare for sea temperature to be above 20°C in temperate to cold climates. Heat loss by respiration also becomes significant as gas density increases. Active clothing, with heated water pumped through tubes in specially fitted suits, is required for prolonged immersion in cold water. Caution is required as the range of safe temperatures of the heated water is narrow.

Nishi and Gagge (1977) consider the heat balance equation and thermal comfort criteria for people in hyper-baric environments and calculate significant increases in convective heat loss due to breathing with pressure. Heat transfer is mainly by conduction and convection, and although principles of human heat balance remain the same, avenues and properties of heat transfer vary from those in air. A fuller discussion of the effects of cold water on diving is provided by Ashcroft (2000) and Hayes (1989). Water has a specific heat, c_p (J kg^{-1} °C^{-1}), thermal conductivity, K (W m^{-1} °C^{-1}) and density, ρ (kg m^{-3}), significantly higher than that of air. This accounts for the high heat loss, by convection in water and, in hyper-baric oxy-helium mixtures used in diving (Hayes, 1989).

Much of the research into human response in water is related to survival times. Swimming pools are kept at around 28 °C, assuming active people (maybe with some solar radiation and wind, outdoors) and would not be expected to cause unacceptable thermal strain. For survival in the water it is best to wear survival suits or as much clothing insulation as possible with a lifejacket for buoyancy and to support breathing. Heat loss in water is so great compared with that in air that it is always better to be out of the water or partially out of it than in it. This can be shown with a thermal audit (Parsons, 2014) as well as empirically. Brooks (2001) demonstrated, in a controlled experiment at sea with the Canadian coastguard, that it is better to stay on an upturned boat, even in wind, rain and 'swell' than stay in the water. Anecdotal evidence confirms this. One of only six fishermen survived, when their vessel sank in cold seas, even with some wearing survival suits and lifejackets. It was because the survivor found a plank of wood to rest on. The Fastnet race off the south coast of the UK in July 1979, when 19 lives were lost in a severe storm, provided the adage: 'always step up into the life-raft'. This was because people who died in the rough cold water had deserted their 'sinking' vessel too early and those that stayed with it tended to survive.

A modification to the open life-raft to include a canopy that prevented exposure to wind probably saved the lives of ditched pilots and seamen who had abandoned their ship. In boats in cold weather, avoidance of hypothermia will depend upon keeping dry and out of the wind. Wearing a lifejacket, being fat, staying hydrated, keeping oneself well insulated with an impermeable outer layer, staying well fed, optimistic and determined and having a high expectation of surviving, along with keeping high morale, sober and more, are all key components of safety at sea. The practical advice for avoiding unacceptable thermal strain can be determined theoretically using the human heat balance equation or a thermal model along with details and consequences of contextual effects. Golden and Tipton (2002) provide further information and practical advice based upon physiological response to cold and theoretical considerations. Only in exceptional circumstances and in exceptional people (insulated by fat) should extensive swimming be contemplated.

UP MOUNTAINS

There is lower atmospheric pressure as we rise above sea level because there is less atmosphere above. Barometric pressure is greater at the equator than at the poles and is highest in summer. There is also lower oxygen content in the air as it is still around 20% but the air is less dense at higher altitudes, so for a fixed volume there is less oxygen. Ashcroft (2000) shows changes in atmospheric pressure from 760 torr (760mmHg = 1013mb=101.3kPa) at sea level through 380 torr at around 5,500 metres as the highest continued human habitation to around 220 torr at 9000 metres around the top of Mount Everest. Passenger aeroplanes compress air to around 2,000 metres with pressure outside at about 10,000 metres. Boiling point decreases with pressure so at the top of Mount Everest water boils at 68 °C and blood (at around 37 °C) boils at around 19,000 metres. The partial pressure of oxygen in the lungs at sea level is around 100 torr and of carbon dioxide it is 40 torr. Actual measurements on Mount Everest show an atmospheric pressure of between 250 and 300 torr, a partial pressure of oxygen in the lungs of around 40 torr and a partial pressure of carbon dioxide around 10 torr. Deeper breathing, more room for oxygen due to a decrease in carbon dioxide, particular meteorological conditions and probably factors not yet understood, have shown that it is possible to reach the summit of Mount Everest and return without supplementary oxygen.

In terms of heat transfer up mountains, there is greater heat loss by evaporation but less heat loss by convection. Athletes training at altitude often think that they are not sweating as sweat evaporates quickly. Solar radiation will be more intense and mechanisms of conduction will remain the same. As we move higher in the atmosphere air temperature will decrease at around 1°C for every 100m climb, although it will depend upon the weather and wind will greatly contribute to cold stress. Surfaces will be cold and will cause damage to exposed skin especially when in contact with metals (see Chapter 12).

Parsons (2014) considers the human heat transfer equation and suggests multiplying the equation for convective loss through breathing by the number of atmospheres (ATM at sea level=1.0; <1.0 up mountains; >1.0 under water, in pressurized tunnels, etc.). For heat transfer by forced convection, multiply the air velocity value by ATM and for heat transfer by evaporation, divide the evaporative heat transfer coefficient for evaporation by ATM. Metabolic heat production will be related to activity level but the capacity to carry out work will be greatly affected by the oxygen available at the cells. At high altitudes, capacity to carry out work will be greatly reduced and hence metabolic heat production will also be reduced, increasing the cold stress.

UNDER PRESSURE

Evaporative heat loss will be greatly reduced in high-pressure environments and as sweating under cold stress is less likely this will have reduced effect. Convective heat loss will greatly increase in high-pressure environments. In hyper-baric lifeboats (e.g. for divers) and in tunnels and caissons (used to keep out water in construction or for repair), high pressure will reduce the margin of temperatures that are safe and acceptable before unacceptable heat and cold strain occur. Similar equations for the assessment of heat transfer to and from the body provided by Parsons (2014) above apply but for ATM now > 1.0. Fuller discussions are provided in Hayes (1989), Parsons (2014) and Ashcroft (2000).

IN SPACE

Space has a temperature of about 2 K (–271 °C) and is mostly empty (even with the presence of dark matter). People in space must therefore be protected from extreme cold and when the sun shines in their direction from extreme heat. This is achieved through sophisticated space-suits, space-ships and orbiting laboratories. It will not be long before we have habitable buildings on other planets (probably first on Earth's moon). Space also has zero gravity and planets have different gravities depending upon their mass. The principles of heat transfer still apply, however, and the specific contexts will modify formulae. Heat transfer by convection for example will have no or modified natural convection depending upon level of gravity.

There is a great deal of research that needs to be done to determine human physiological response to cold environments experienced in space and beyond. This will be required along with many technological discoveries and innovations to be able to build a comprehensive understanding of responses of people in extra-terrestrial environments.

IN THE ARCTIC AND ANTARCTIC

The coldest place on Earth, if we use air temperature as the measure, was recorded −89 °C (−129 °F), at the Soviet Vostok ice station in the Antarctic in July 1983. The Arctic is made of ice and the Antarctic is made of land and ice. The sea around them freezes at −2°C and human flesh freezes at −0.5 °C, so best not to leave naked skin in the water for too long. People who live in the polar regions, explorers and those on scientific surveys, mainly survive the cold by avoiding it. Wearing heavy clothing for long periods can become depressing and, as with space, self-selected experienced individuals are those mainly exposed (Conway et al., 1998). There have been many expeditions to the Arctic and Antarctic and as with mountaineering and space, a pioneering spirit with a tinge of commercialism will drive people on. The epic journeys and experiences of Scott, Shackleton, Amundson, Peary and more, as well as the scientific work of Siple and others are well documented. Siple (1945) provides an excellent practical description of clothing for cold climates and this was followed by the classical influential work of Siple and Passel (1945), who reported on the work of the United States Antarctic Service in 1941. It is encouraging to know that scientific knowledge has progressed so much to the present day that the principles described in this book on human cold stress can be used to build upon the previous work to provide practical advice for all cold environments experienced on Earth and beyond.

SPORTING ENVIRONMENTS

People involved in active sport usually sweat, have metabolic rates much greater than when at rest and have a high relative air velocity due to movement. They will become cold and chilled in cold conditions when they are at rest so recovery procedures (e.g. increasing clothing insulation or going to a warm area) are required. Armstrong (2000) asks us to imagine going on a jog on a cold day in light clothing. Skin temperature falls immediately and vaso-constriction follows. Stroke volume increases due to venous return to the heart and cardiac output and heart rate are lower than they would otherwise be in warmer environments. Muscle temperatures fall (from initial warm-up values), promoting anaerobic metabolism.

The Education and School Premises Regulations (1999) in the UK states that it should be possible to heat classrooms to 18°C and that gymnasiums should be at least 15 °C. Environmental limits for sport outside in the cold will depend upon wind chill. This will depend upon the particular sport (ACSM, 1984). Skiers will normally be protected from the cold across all body parts due to clothing. ISO 11079 (2007) provides physiological criteria that can be used as a basis for limits for sporting events in the cold. A wind chill tempera-ture (twc – see Chapter 9) of –30 °C provides a lower limit for sport with a limit for air temperature of –30 °C for breathing if activity level is high.

Diverse Populations in the Cold

12

DIVERSE POPULATIONS

DIVERSE POPULATIONS

Being cold is universally unpleasant and there are similar physiological mechanisms for responding to cold stress among all populations. Newborn babies and babies in general are vulnerable because of their immature development, large surface area to volume ratio (efficient for heat loss) and a tendency to be wet. In the UK it is mainly babies and old people that are admitted to hospital with hypothermia (Collins, 1983) and the elderly and the vulnerable who die when the weather turns cold.

Body size and constitution are of great importance. For similar conditions, fat people will survive for longer in extreme environments than thin people (especially in cold water). Small noses, tucked back small ears and fat stubby fingers will reduce heat loss. Psychological considerations suggest that the will to tolerate and survive in the cold and refusing to give up has a significant effect but the most important individual characteristic in combatting cold is human behaviour driven by discomfort, or the perceived threat of discomfort, derived from experience and culture.

BABIES

Babies have relatively large heads and surface areas for heat exchange as well as high blood flow and limited capacity to sweat. Their skin has a high water

93

content, a thin epidermis and newborn babies are wet, all promoting heat loss by evaporation. Clarke and Edholm (1985) suggest a thermo-neutral incubator temperature for naked newborn babies as 32 °C and note that as a baby becomes cold, structures under the skin become visible with infra-red cameras. Collins (1983) notes that the foetus 'in utra' temperature is around 0.5 °C higher than that of the mother, that metabolic heat is produced by the foetus and that any heat exchange is through the placenta. The newborn baby is greeted with sudden cold, light, noise and more. Preserving core temperature between 35.5 °C and 37.5 °C is optimum for growth and development. Hypothermia can have dangerous consequences and premature babies are particularly susceptible. It is exacerbated by relatively low basal metabolic rates (per surface area) when compared with those of adults (although higher per unit body mass).

Babies can vasodilate when hot and vasoconstrict when cold. Most babies exhibit thermogenesis and maybe shivering when cold, and brown fat provides extra heat in the cold. It is located in the neck, spine and around vital organs and contains numerous blood vessels and an increased density of mitochondria.

Babies are not capable of effective behavioural thermoregulation and rely on 'carers' to correctly perceive their requirements and act upon them. Hypothermic babies show red cheeks, due to an inability of oxy-haemoglobin to disassociate its oxygen to cells, skin cold to the touch and a low body temperature (usually with an obvious environmental cause). Signs of illness are lethargy, difficulty in feeding, slow heart and respiration rate and a cold body. Swollen limbs and face are more advanced symptoms requiring hospital treatment and careful re-warming.

CHILDREN

On standing up, toddlers expose most of their surface area to the air and up to the adolescent growth 'spurt' their body surface area to volume ratios are conducive to heat loss and usually greater than that of adults. Small thin children with thin fingers and toes are particularly susceptible to cold.

The metabolic rate of children requires further research; however, basal metabolic rate per unit of body mass is generally higher than that of adults and children are more active than adults. Physiological thermoregulation tends towards that of adults; however, behavioural thermoregulation is often naive and inappropriate. Children require guidance and supervision to ensure effective behavioural thermoregulation in the cold.

ELDERLY PEOPLE

As people become old, experience of how to behave under cold stress consolidates but their physiological and behavioural capabilities decrease. Physical and mental fitness is an important factor but, in general, susceptibility to cold stress increases with age. There are many individual differences as different parts of the thermoregulatory system deteriorate in different people at different rates. There is, however, a general decrease in metabolic rate and reduction in the level of activity with age.

Collins (1983) suggests that the loss in thermoregulatory capacity, due to inefficiency, disease and drugs (that interfere with normal thermoregulation) as well as the inability to tolerate extreme conditions, poses a challenge to maintain internal body temperature at optimum levels in the elderly. He notes that metabolic rate is reduced in the elderly, possibly influenced by the lower proportion of functional cells in the body; that surface area to volume ratios alter with age (especially in the obese or 'frail'); that core temperatures are around 0.2°C lower in the elderly when compared with younger adults; and that skin temperatures may be higher in some elderly people, causing greater heat loss in the cold, due to restricted vasoconstriction in some (around 20%) individuals.

Shivering is thought to be greatly reduced in the elderly but (as with younger adults) there are large individual differences. As hypothermia is approached, however, shivering occurs in both young and old. Temperature sensation is reduced in older people that can reduce discomfort and the drive to restore comfort through behaviour. Collins and Hoinville (1980) consider temperature requirements for people in old age and Coull (2019) present body maps of thermoregulatory responses and ageing.

As people become old, their physiological systems in general deteriorate as does their ability to take behavioural action. Cerebrovascular diseases and diseases of the heart and other organs will restrict capability. Limitations in movement or cognitive dysfunction, may commit elderly people to cold exposures as they are not capable of moving away. Hypothyroid diseases can reduce metabolic rate and internal heat production to such an extent that it promotes hypothermia. Rate of change of temperature can cause dramatic changes in blood pressure (hypertension is a common and medicated disease throughout the world) that can cause injury and death. Tochihara et al (1998) showed a connection for elderly Japanese people in cold bathrooms entering hot baths and it is not advisable for the vulnerable and elderly to dive into cold water after a hot bath or sauna.

The nature of deterioration in thermoregulatory response to cold stress is widely recognised but requires further research to provide more detail of individual responses. There is a general belief that cold showers, cold baths and 'wild' swimming, as well as immersion in cold and icy water, are beneficial. There is some evidence regarding an improved immune system, and psychological benefits of experiencing stress are clear. Cold can also damage the health of the vulnerable and the elderly, however, especially if extreme, and further research is required.

GENDER

It is common experience that females in offices suffer greater cold discomfort than males (mainly due to cold hands because of small fingers but it may also be confounded by clothing). Female clothing often has greater areas of exposed skin and capacity for ventilation. Gerrett et al (2015), in their body mapping studies, found greater sensitivity across the body, of females than males to a cold stimulus. Brager et al (2004) reported greater adaptive responses of females than males in offices although as well as sensitivity to thermal conditions, this could be related to psychological and cultural factors (see also Hashiguchi et al, 2010).

Despite anecdotal 'evidence' there appears to be no actual evidence of the effects of the menstrual cycle (Fanger, 1970) or of the menopause on responses to cold. There is also no evidence of gender effects of circadian rhythm on responses to cold. Further research is needed into those factors. We should remind ourselves at this point that cold stress is the interaction of air temperature, radiant temperature, humidity and air velocity in the environment as well as the personal factors of clothing and activity level. So for example, in that context, less clothing and greater cold stress is not a gender effect.

There appear to have been no systematic studies into the effects of gender on responses to very cold and extremely cold conditions. Clearly magnitude of physical strength will show differences but any differences in responses to severe cold stress have yet to be demonstrated. When I investigated worker performance in a freezer room at −23 °C and in an interview with the shift manager, he indicated that he employed workers of both genders and found them to be equally effective. Some caution is required as shift working particularly in freezer rooms is a self-selecting activity often avoided by older workers.

PEOPLE WITH DISABILITIES

People with disabilities are particularly vulnerable to environments that deviate from optimum due to their particular disability, the influence of any drugs they take, limitations in their physiological response, restrictions placed upon them by technological aids and, most importantly, their ability to behave in a way to reduce stress and avoid unacceptable strain (Reegan 2017). All of these factors apply to cold stress. A person who cannot adjust clothing or move away from even a mild cold stress will be in danger and find it unacceptable and even frightening.

Parsons (2020) presents a discussion and data based upon the results of Webb and Parsons (1997). The studies were to investigate thermal comfort requirements for people with physical disabilities. They found that even within those 'labelled' with the same disability, there were a wide range of responses to thermal environments. It was concluded that people should be considered as individuals with their own specific requirements if optimum conditions were to be achieved. In laboratory studies and studies involving homes and care homes (a total of 568 people with physical disabilities and 38 carers) a general conclusion was that people with disabilities tended to want to be warmer and avoid being cold. Physical disabilities studied included spinal injury, spinal degeneration, spina bifida, hemiplegia, polio, osteoarthritis, rheumatoid arthritis and multiple sclerosis (ISO 22411, 2021; Parsons and Webb, 1999; Parsons and Webb, 2013; Webb and Parsons, 1997; Webb et al, 1999).

The climatic chamber studies included an environment at 18.5 °C, which would be considered on average, slightly cool to cool, by people without disabilities (PMV = -1.5, ISO 7730, 2005).

It was found that people with hemiplegia (including stroke, N = 4) and spinal degeneration (N = 8) tended to feel cold. People with disabilities in care homes generally wanted temperatures to be higher. Parsons (2020) notes that none of the people with disabilities of all types in care homes wished to be cooler. For cold conditions, adaptive opportunities should be emphasized. Restrictions to adaptive opportunities are the ability to move and move around, adjust clothing, rely on assistance or use technical aids. Even when adaptive opportunities are potentially available, people with cognitive impairments and limitations may not be able to take advantage of them.

People with disabilities who have restricted vasoconstriction, large surface areas for heat loss and restricted thermogenesis and shivering due to drugs, metal inserts, and restricted ability to behave appropriately, will justifiably find exposure to cold unpleasant and threatening. Sitting in a draught without the ability to move away is an undesirable state. When exposed to cold, people with disabilities will therefore probably become uncomfortable and eventually their internal body temperature will fall to hypothermia unless careful consideration is given to both physiological and behavioural requirements. Disability discrimination legislation, adopted internationally, requires that people with disabilities should not be discriminated against when at work. Careful attention to the effects of cold will be required to meet with such legislation.

Yoshida et al (1993) noted that people with disabilities often experienced overcooling due to disorders in peripheral blood flow. International research in this area was integrated into ISO TR 22411 (2021) and for responses to physical environments in general, ISO 28803 (2012).

PREGNANT WOMEN

If a pregnant mother has hypothermia (or a fever) the foetus will suffer the same as all heat exchange is via the placenta mainly through blood supply. Parsons (2020) suggests that no difference has been shown between thermal comfort conditions for non-pregnant and pregnant women at rest, although research is required. Pregnancy is not of one form, typically lasting for 40 weeks and with great changes in hormonal, morphological and other characteristics. An active pregnant woman in advanced stages of pregnancy has limited adaptive opportunity and increased metabolic rate when moving due to increased weight and weight distribution.

Clearly, many babies are born in cold climates and cultural knowledge will ensure that the mother and foetus are kept warm. For severe hypothermia it will be important not to assume that either mother or foetus have died as vital signs will be difficult to detect. Careful re-warming under medical supervision will be important. More research is needed into the effects of cold on pregnant women.

PEOPLE WHO ARE ILL

Cold stress can directly cause illness, hypothermia and injury to a person (see Chapter 13). It can also affect people who are already ill. The additional strain will generally make things worse and care must be taken in hospitals to ensure that sedentary patients do not become cold.

For people with extreme fever, where erratic behaviour and delirium can lead to severe consequences, reducing body temperature with cold (cold water bath) can have advantage if done with care. Fever raises body temperature, blood flow and skin temperature, so exposure to cold promotes significant heat loss. Cold environments can sometimes offer advantage by slowing metabolic rate and hence the progression of an illness. Collins (1983) provides a full discussion.

Overall no general recommendation can be made to provide cold conditions for people who are ill. It is reasonable to assume that cold conditions will provide an additional strain on the body, including the immune system, and complications such as pneumonia may be caused. With no additional information, avoiding hot and cold conditions and planning for thermal comfort and reduced strain seems a logical approach to environmental design (Parsons, 2019, 2020).

PEOPLE INVOLVED IN SPORTS, WALKS AND FUN RUNS

People involved in sporting activities, including those for leisure, usually test their physiological systems to the limit. They generate high levels of metabolic heat and sweat, are highly motivated, are clothed for the activity (so inappropriately clothed when they stop) and feel exhilaration a deadly combination in extreme cold. Castellani and Tipton (2015) and Castellani et al (2010) provide reviews. Older people are particularly vulnerable and may not benefit from a walk on a cold day (Keatinge and Donaldson, 1998).

Taking a 'fun' run in the cold especially when not used to it is also not advised. Rate of drop in temperature is important and a rapid drop in skin temperature (particularly the face) such as when turning to the wind or when diving into cold water can greatly increase blood pressure. The adage "what doesn't kill you makes you stronger" seems to be particularly relevant. "The situation for the runner in cold weather is ... massive heat loss caused by vasodilatation, sweating and a wind effect on the body surface ... offset by a high rate of internal heat production" Collins (1983). He notes that if the runner is forced to stop (exhaustion or dehydration) then hypothermia is a great risk.

Castellani et al (2006) provide a position stand for the American College of Sports Medicine for the prevention of cold injuries during exercise. They recommend a comprehensive risk assessment strategy. Full training and a plan of action is required and they note that at wind chill temperatures of below

−27 °C (−18 °F) there is a heightened risk of frostbite. People with asthma and cardio-vascular diseases should be monitored closely.

Collins (1983) provides a comprehensive review of accidental hypothermia due to cold outdoor weather conditions with examples of organised activities that have led to hypothermia and death. Hypothermia is difficult to identify in its early stages and diagnosis should not be left to those exposed. Motivation and extreme sport are a dangerous combination. Exhaustion often gives way to continuous shivering and slowing, and loss of coordination and orientation are common, leading to confusion. Rapid rewarming is necessary and if a hospital or other sanctuary is unavailable, rest in a sheltered area is recommended. If clothing is wet and can't be replaced, cover with an impermeable layer. To retain water in clothes and heat one litre from 4°C to 34 °C in one hour requires 35 Watts of heat, whereas one litre evaporated from the skin would require 675 W of body heat (Collins, 1983). Other advice includes use of a casualty bag, covering the head, mouth and nose with a scarf, body movement if not too impaired (beware of massaging the skin as it may return cold blood to the heart), body contact with others, surface warming and use of heat sources such as a fire, hot sugared liquids and avoidance of alcohol (St Bernard dogs carried sugared liquid only lightly laced with brandy). Importantly, morale, involvement, discourse, singing, encouragement, conversation and hope must be maintained. I was asked how to rewarm rugby players who had become cold (not hypothermic) on the side-lines as potential substitutes. An answer is to exercise (on bicycle ergometers out of the wind and preferably in a changing room), avoiding sweating, but to rewarm with metabolic heat and blood circulation.

A major development, beneficial to individuals and to society, has been the participation of people with disabilities in sport. The principles of human response to the cold in general combined with the knowledge presented above for people with disabilities, and specific to individual competitors, will provide a starting point to provide advice for ensuring safe environments. More systematic research is needed in this area, with particular attention paid to the experiences of people with disabilities involved in sport.

From running to ski-ing and other activities, relative air movement across the body will be important and can cause significant cooling especially to the hands and face. Running into the wind is additive and running in the same direction as the wind at the same speed provides relatively still air on average across the person. However, hands move backwards and forwards, air can be turbulent, people change direction and so on.

Whatever the sport or activity, the principles of human response to cold apply, and the air temperature, radiant temperature (greatly influenced by the sun), humidity, air velocity, clothing and activity will determine cold stress and provide the rationale for exposure limits, avoidance of unacceptable cold strain and management systems of people and groups exposed to cold.

All people are exposed at some time to cold environments of many types. These vary from homogeneous cold stores to a wide range of diverse, including outdoor, environments often dictated by the weather on Earth but also beyond.

ACCLIMATISATION TO COLD

Some people are said to be acclimatised to cold and it is a topic of much debate. There is clear evidence of adaptation involving instinctive, learned and cultural responses and behaviour to reduce cold stress. Physiological adaptation (such as clear increases in sweat rates when exposed for prolonged periods to exposed heat) has not been demonstrated for exposure to cold. A fuller discussion is provided in Rivolier et al (1988), Gordon et al (2019) and Castellani et al (2016). There has been some suggestion that people who work with their hands in the cold such as fishermen, can maintain a high blood flow and hence maintain manual dexterity. There is, however, some debate about whether this is simply an indication of damaged hands (Parsons 2014).

Hypothermia, Freezing and Non-Freezing Cold Injuries and Death

13

HYPOTHERMIA

If heat loss from the body is greater than metabolic heat production combined with any heat inputs from the environment, then body temperature will fall. If the internal body temperature (often assumed to be rectal temperature) falls below 35 °C then this is termed clinical hypothermia. It should be emphasised that this is a dangerous condition and that any drop in internal body temperature elicits severe reactions from the body. These include vasoconstriction, shivering and loss of strength, leading to lowering of skin temperatures with associated stiffness and numbness as well as a drop in physical and mental capability. As well as a strain on the heart this may lead to a significant drop in physical and mental capacity and changes in mood with severe discomfort, distraction and behavioural reactions that can lead to accidents as well as inappropriate behaviour and decision making.

Virus activity is said to increase in cold tissues although logically this is probably relative to the reduction in metabolic rate, slower reactions in living cells and in the immune system and general reduced chemical activity at lower temperatures. Hunger and dehydration exacerbate the problem.

Symptoms of hypothermia start with cold skin and shivering with a loss in manual dexterity and cognitive ability. This leads to confusion (taking off clothes, dump rucksack, etc.), lethargy, apathy and loss of memory. As the internal body temperature drops below 32°C (or above in some individuals) collapse can occur as the person 'fades away' and the heart rate and blood pressure reduce towards fibrillation and eventually zero.

An attempt to determine limits of hypothermia after which death will occur was made in the experiments held in Dachau, Germany, in the 1940s. The experiments lacked both validity and morality. Ventricular fibrillation leading to death occurred in anxious, forlorn, weak and vulnerable people immersed in cold water when internal body temperature reached 28°C. This is sometimes taken as a lower extreme; however, it is a mistaken and dangerous interpretation and any drop in internal body temperature will lead to a strain on the body, particularly the heart, and survival will depend upon the health and vulnerability of the individual as well as context and method of recovery.

REWARMING THE BODY

The United States Environmental Protection Agency, EPA (2017), provides guidance on management and emergency responder advice for exposure to physical stresses including cold. The lowest reported internal body temperatures with recovery are around 9°C–13.7°C, depending upon confidence in accuracy of measurement and area of the body measured. These are levels much lower than expected recovery but the message must be 'don't give up'. Death can occur at levels above internal body temperatures of 30°C depending upon the vulnerability of the individual exposed but at low internal body temperatures, reduced metabolic rates and hence oxygen required in cold tissues (e.g. the brain) can aid survival particularly in children and relatively rapid cooling.

Attempts at rewarming shall not be given up. There is clear potential for incorrect diagnosis of death. Let the hospital decide "nobody is dead unless they are warm and dead" (Ashcroft, 2000). Rewarming is mainly about raising the internal body temperature back to normal while avoiding excessive cardiovascular strain including debilitating changes in blood pressure. For mild hypothermia (core temperature >34°C) Golden and Tipton (2002) recommend immersion in a hot bath, at 40°C (Keatinge, 1969). A variation of the method for cooling hyper-thermic casualties (Parsons, 2019) may be useful for rewarming, where the patient is immersed in a warm to normal bath to avoid a rapid change in temperature and hence blood pressure. For a hypothermic patient, the bath is then heated with flowing water at 40 °C until the stirred bath temperature reaches 40°C. 'Filling' half a normal bath and adding water or using a combination of tap and plug-hole are practical methods. Core temperature, blood pressure and heart rate (as well as heart rate variability or even ECG if possible) should be monitored.

Detail of how to rewarm casualties after immersion in cold water is presented by Golden and Tipton (2002). Rewarming of casualties, including those from accidental hypothermia caused by extreme environments when engaging in outdoor activities, is considered by Collins (1983) in his book *Hypothermia*. Rapid changes in temperature and pressure (e.g. when raised from cold water) cause rapid changes in blood pressure and blood redistribution that can be fatal.

Golden and Tipton (2002) discuss the 'after-drop' (continued core temperature drop immediately after rewarming) in this context, where after cold exposure the internal body temperature (rectal temperature) continues to fall (by up to 0.5°C). The causes of the after drop were investigated by Golden and Hervey (1977, 1981) when a pig was raised from cold water immediately after its heart had been stopped. There was therefore no redistribution of blood but still allowing conductive heat transfer through tissues. The outcome was that some after drop was detected. Golden and Tipton (2002) note that any after drop (continual cooling) is seen in rectal temperature (greatly influenced by conductive cooling) but not in the heart and that recovery body posture is important in post immersion recovery. They speculate that hormonal changes as the person in the cold water gives over control to the rescuers, with associated psychological relief, may incur vulnerability. It is clear that the context and nature of hypothermia should influence re-warming strategies. The avoidance of cardiovascular strain is the most important objective for re-warming which if successful will lead to a complete recovery of the patient.

Collins (1983) notes that the appropriate treatment (or method of prevention) of hypothermia will depend upon the individual person and circumstances. Fit, young and healthy people have a much better chance of survival than the elderly and those with underlying medical conditions. Breathing indicates a heart rate so no resuscitation, but maybe defibrillation, is required. For re-warming, monitoring rectal or urine temperatures is useful but 'low reading' thermometers may not be available. Rapid recovery is achieved through a hot bath (41 °C to 45 °C) with limbs out of water to reduce any after drop. For slower rewarming hypothermic babies are rewarmed in incubators and very warm rooms. Adults benefit from showering and are revived actively with warm blankets and covered (to avoid burning) hot water bottles as well as warm baths and warm water-filled mattresses. Internal heating can be achieved through transfusions of warm blood with additional marginal advantages of warm drinks and food as well as breathing warm air. Passive rewarming relies on significant clothing, blankets and metabolic heat. It is recommended that elderly people are rewarmed slowly to avoid rapid change and consequent shock to the body (see Collins (1983) and Pozos and Danzl (2001) for a full discussion).

COLD INJURIES

Damage to the body due to cold can be divided into non-freezing cold injuries, mainly manifesting in the foot, also called immersion foot in water and trench foot on land (often experienced in military campaigns), and freezing cold injuries mainly due to the freezing of the cells leading to frostbite Hamlet (1988, 1997). The effects of temperature on water play a dominant role. The fluids in cells freeze below −0.5 °C, frozen fluids expand and the cell wall is penetrated, the cell is dehydrated and the complete cell is irreversibly destroyed. Sea water freezes at around −2 °C so potentially cells can freeze in liquid sea water.

NON-FREEZING COLD INJURIES

Exposure to cold can cause damage to moist skin, particularly the lips, and balms are important. Chilblains are red and 'itchy' patches of skin, in extremities such as the hands and feet, experienced during rewarming and are a result of repeated exposure to cold. Raynaud's syndrome is caused by severe vasoconstriction of extremity blood vessels in the cold followed by a painful rewarming. It is exacerbated by vibration of the hands, with associated damage to blood vessels and nerves (Griffin, 1990).

Golden and Tipton (2002) notes that in cold and wet conditions, the muscle pump that pushes blood 'uphill' back to the heart ceases to operate and swelling occurs aided by blood pooling and maybe tight boots as well as vasoconstriction due to anxiety. Trench foot is caused by wet feet with water temperatures below around 10 °C (Golden and Tipton (2002) suggest even up to 17 °C). Vasoconstriction and reduced circulation of blood reduces available oxygen, causes numbness and a pale mottled colour, often leading to cell death sometimes first in internal tissues so with no external signs. Severe cases lead to amputation of the foot, and the degree of debilitation among troops, due to trench foot, can decide the outcome of military operations with many examples up to the present day. Recent examples include the Glastonbury pop festival in the mud of the UK summer as an example of the occurrence of significant numbers of civilian cases (Ashcroft 2000). Golden and Tipton (2002) provide numerous practical accounts of immersion foot where sea-farers have exposed their feet to immersion in cold water for long periods. As well as deaths, amputation of toes and long-term reduced mobility are not uncommon.

FREEZING COLD INJURIES

Frostnip is detected as a sharp pain and leads to freezing skin with a waxy appearance, mainly in the fingers, toes, nose, ears and cheeks often caused by wind chill. 'Superficial' frostbite freezes skin and underlying tissues and is severe with large blisters, but there is a possibility of recovery. Don't take gloves off when handling cold materials! More profound frostbite freezes skin as well as internal tissues such as muscles, bone and tendons and usually leads to amputation (mostly of fingers, toes and even feet and hands). For both non-freezing and freezing cold injuries additional symptoms can include skin ulcers, gangrene, oedema and more. I have spared the reader photographs of withered hands and black, apparently charred, frostbitten fingers. For a fuller review, see Oakley (1990) and Oakley and Loyd (1990). Goldman (1994) suggests that the Wind Chill Index of Siple and Passel (1945) was found to be valid for predicting when fingers would freeze, during unacceptable and immoral experiments on US prisoners of war in Manchuria during the Second World War (see also Hanson and Goldman, 1969; Wilson and Goldman, 1970 and van Dilla et al., 1957).

PSYCHOLOGICAL 'ILLNESS'

Had Scott and his team of explorers been the first people to reach the South Pole they would have returned with a spring in their step and a positive attitude excited to return to a welcoming public and glory. Instead Amundson and his team were first and they returned to the glory that Scott had created and a despondent Scott and his colleagues died. This is speculation, but it is well recognised that good spirits and a will to survive are psychological traits that aid survival.

Rivolier et al (1988) describe the psychological effects of living in tents over a winter in the Antarctic to study possible acclimatisation to cold and other effects. Being cold was not a major problem as clothing provided protection, but living in the cold promoted significant psychological responses from anxiety, introspection and loss of confidence to significant social problems with a breakdown of social relationships and group cohesion.

STATISTICS

Winter excess deaths in the UK are measured by the Office for National Statistics and were a measure of the death rate from January to March when compared with the average for the whole year. More recently excess winter mortality is determined by comparing the winter months of December to March with the average of the four-month periods before and after. There are tens of thousands of excess deaths every year (28,300 in England and Wales for Winter 2019–2020, excluding deaths related to the COVID-19 pandemic) and it is reasonable to infer that this is related to additional cold stress with respiratory illness the main cause. A possible starting statistical model for investigating excess deaths would be to consider total excess deaths as those caused by main effects of Cold and 'Covid' plus a Cold × Covid interaction. Without the cold there is only Covid and no interaction.

Statistical 'evidence' suggests that cold stress is linked to illness and death mainly among the very young, the elderly and people with underlying medical conditions. Staying indoors in cold weather can promote infection among occupants and outdoors it can provide unacceptable strain on the vulnerable. The use of urine temperature in surveys of old people (in place of unreliable measurement of oral temperature) improved validity of results and showed that mild hypothermia was common in the morning among elderly people in the UK. Indoor central heating (and financial support for heating bills) has reduced seasonal excess deaths. Pandemics such as that caused by the COVID-19 virus have greatly increased the number of excess deaths in 2020–2021 (so far in the UK to over 125,000, mostly elderly people) with a strong indication that cold conditions have promoted infection and severity. The causal mechanisms remain to be investigated.

BENEFITS OF COLD

The psychological benefits of exposure to cold for pleasure are well known, from ice plunges after saunas to winter swimming often involving stimulating senses with shifting cold-warm exposures and breathing fresh cold mountain air. They provide stress to the body and can be stimulating but, by definition, strain the body systems (e.g. blood pressure and heart) and can be dangerous where they are beyond the tolerance of individuals.

Collins (1983) lists some benefits of cold: preventing pain; aiding heart surgery; preserving human cells (blood, sperm, eggs), tissues (corneas, cartilage, smooth muscle) and organs; and reducing inflammation and blood flow.

Long-term storage of organs or whole bodies by deep freezing with recovery, has not been successful, and for people cryogenically frozen after death, DNA techniques and the use of STEM cells may show promise, but resuscitation after death seems unlikely.

Skin Contact with Cold Surfaces

14

COLD, PAINFUL, NUMB AND FREEZING

When warm skin comes into contact with a cold surface, heat will be lost from the skin and the skin temperature will fall. If the skin temperature is reduced to below −0.5 °C then the skin will freeze and be damaged. Above −0.5 °C the skin will range from uncomfortably cold to painful and numb.

Figure 14.1 shows a solid surface coming into contact with skin. For short exposures or cold but not freezing temperatures, the epidermis is cooled, causing discomfort, pain and numbness as sensors are affected. For longer exposures and lower temperatures, the dermis and tissues underneath cool. For freezing temperatures frost nip damages cells on the skin surface and frostbite becomes more profound as both epidermis and dermis freeze, damage cells and lead to extreme injury.

A SIMPLE HEAT TRANSFER MODEL

Skin temperatures which cause discomfort leading to pain and damage are provided in Parsons (2014) and in ISO 7730 (2005) for discomfort and ISO 13732-3 (2005) for skin responses to contact with cold surfaces. A first approximation calculation of the temperature of the surface of the skin when in contact with a cold object can be derived from a simple model of heat transfer.

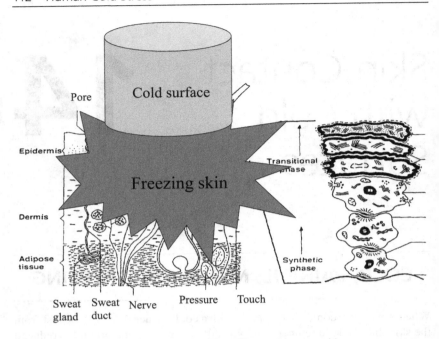

FIGURE 14.1 Contact between a cold surface and human skin.

When two objects come into contact, the second law of thermodynamics states that heat will flow from the object at a higher temperature to the object at a lower temperature. Fourier's law dictates that the rate of flow will depend upon the temperature difference between the objects and that, all other things being equal, the two objects will eventually reach the same intermediate temperature. The 'zeroth' law of thermodynamics states that 'when two objects of the same temperature come into contact there will be no net heat flow'.

For two slabs of material of infinite capacity in perfect contact, the temperature of each slab will remain constant and the rate of flow will be at a steady state. The temperature between the two slabs (1 and 2) will therefore be constant and is referred to as the contact temperature (tc). For the simple model described $tc = (b_1t_1 + b_2t_2)/(b_1 + b_2)$, where $b = (k\rho c)^{1/2}$ and t is the temperature of the material; k is the thermal conductivity of the material; ρ is the density of the material; and c is the specific heat capacity of the material.

For a person taking gloves off to hold a metal peg at −25 °C (especially in the wind that will destroy any boundary layer) with the skin temperature at 30 °C and thermal penetration coefficients for the skin and metal at 1,000 and 20,000 Jm^{-2} $s^{-1/2}$ K^{-1} respectively (Parsons, 2014) gives a contact temperature and hence an estimated skin temperature of (30 × 1000 + −25 × 20000)/21,000 = −22.4 °C. ISO 3732-3 (2005) indicates that contact with aluminium at −22.4 °C would provide 'instant' numbness with pain and progressive frostbite for instant exposures expected to increase in severity to 1 second and beyond.

ISO 13732-3 (2005). ERGONOMICS OF THE THERMAL ENVIRONMENT – METHODS FOR THE ASSESSMENT OF HUMAN RESPONSES TO CONTACT WITH SURFACES. PART 3: COLD SURFACES

International standards concerned with human skin contact with hot, moderate and cold surfaces are presented in ISO 13732 parts 1, 2 and 3, respectively (see Parsons, 2014). Part 3 presents surface temperatures that, on contact with a cold surface, would cause pain, numbness and injury to human skin, particularly but not exclusively on the hands and fingers. Contact is divided into finger touching (up to 120s of contact) and hand gripping (up to 1200s of contact).

A drop in contact temperature to around 15 °C causes pain, to 7 °C, numbness and to below 0 °C, frostbite, when contact tissue freezes. For each of the effects, graphs of surface temperature (not to be confused with contact temperature) vs contact time (period) are provided. For finger contact with metals and a 20 s exposure, pain occurs at surface temperatures towards 15 °C; numbness around 5 °C; and frostbite −4 °C tending towards 0 °C for longer exposures. For contact with a nylon surface, and a 20s exposure, pain will occur at 1 °C and numbness at −33°C. For wood and 20s contact time, pain occurs at −4 °C and numbness below −40 °C. For both wood and nylon, frostbite thresholds are not specified and can be assumed to occur well below −40 °C. For hand gripping it would take 100s to reach a pain threshold for aluminium at −4 °C and 200s for wood at −30 °C. ISO 13732-3 (2005) provides thresholds for aluminium, steel, stone, nylon and wood as well as material properties that can be used to determine contact temperatures. The international standard also provides risk assessment methods and practical advice on how to determine the likely effects of skin contact with cold surfaces.

Sticking

If the skin is wet or moist or the surface is wet, and if contact temperature is below 0°C then the skin will stick to the surface. Almost without exception my masters students from Canada remembered, when children, of being warned not to take up the dare of placing their tongues on cold metal surfaces (poles, windowsills, etc.). If they did so they should not pull back but warm up the surfaces until the water melts! Metal surfaces not so far below 0°C will create freezing conditions for moist skin and the skin will stick to the metal. Taking gloves off to work on cold metal objects (tempted by the promise of improved manual dexterity) should be avoided in very cold conditions. Metal objects will cause frostbite and the possibility of sticking.

Practical Assessment of Cold Environments

15

THE SURVEY

Millions of people throughout the world work in cold indoor and outdoor climates, and advice is often required on how to design and assess those environments. A natural and important starting point to practical assessment is to consider human response to thermal environments as being determined by the air temperature, radiant temperature, humidity and air velocity in which the people work as well as their activity and the clothing that they wear. The opportunity afforded to them by the environment to reduce their exposure to cold stress is also of importance (e.g. rest in a warmer area and increase clothing levels).

For existing environments any practical assessment will involve measurement of the environmental conditions, measurement and prediction of how those conditions affect the people who occupy them and advice on how to improve those conditions. The well-being of the people is usually of prime importance as well as completing any tasks. For that, it is usual to consider the effects of cold stress on the health and safety, comfort and performance of the people in the environment. A comprehensive treatment of ergonomics methodologies is provided in Wilson and Sharples (2015). Tochihara (1998) provides a practical example of assessment of cold work.

There are many cold environments to consider including those in and on the water. For environments related to cold water immersion, survival is usually the main objective and comprehensive coverage is provided by Golden and Tipton (2002). For any assessment of cold environments the principles of assessment are similar. Cold environments are sometimes defined as those where the air temperature is less than 10 °C but this is arbitrary (people can be hot below 10 °C and cold above it). An example of the assessment of a cool to cold environment is provided below. Principles and guidance for conducting an Environmental Ergonomics Survey are provided in ISO 28802 (2012) and Parsons (2005a, 2005b).

A primary consideration when conducting a practical assessment of a cold environment is to determine where, when and what to measure. It is of great importance to determine at the outset, exactly what the aim of the survey will be and what will be delivered. Where to measure will preferably be at the places where people will be exposed to the cold stress. This is a non-trivial point as the systems for controlling temperature are often taken as representative of the cold stress experienced by the people; however, measuring instruments for that purpose are often placed away from people, on a wall for example. The objective is that the measurement of cold stress will be representative of that to which people are exposed. Measurement at workplaces is preferred but a grid system is often used for large spaces, with large numbers of people particularly for dynamic work where people move around. The physical representation of the workplace is termed 'cold stress' and the response of people is the cold strain.

MEASUREMENT OF COLD STRESS

Cold stress is defined by the integration of the effects of air temperature, radiant temperature, humidity and air velocity along with the heat produced by the activity of the person and the insulation provided by clothing. The opportunity to influence the cold stress experienced by people is related to cold stress but any behavioural response will be regarded as cold strain.

For measurement of very cold environments I have found that cold can affect the measuring instruments. Wires stiffen and batteries lose power often with consequent loss in display function and possible errors in measurement. Loss in experimenter performance (e.g. due to cold, distraction and protective clothing), or even safety, should not be ignored. Calibration of instruments and measurement practices should all be determined with respect to the cold environment within which instruments and operator will be performing.

Air temperature (ta) is the temperature of the air that is representative of that which drives the convective heat transfer from the body to the environment. It should not be measured too close to people so that they influence the measurement and it should not be so far away from the workplace that it is not representative. It is measured with thermistors or other transducers (shielded from radiation if necessary) that provide an electrical signal. It is also measured with the dry bulb of a whirling hygrometer that is rotated in air to negate radiation effects. The use of mercury in glass thermometers is not recommended as breakage can release dangerous mercury into the environment. ISO 7726 (1998) recommends a sensor accuracy of ± 0.5 °C for temperatures above 0 °C and $\pm(0.5 - 0.01ta)$ °C for temperatures below 0 °C down to -40 °C with a response time as short as possible.

Mean radiant temperature (tr) is the temperature of a uniform enclosure with which a small black sphere at the test point would have the same radiation exchange as it would have with the 'actual' environment. In other words it is a three-dimensional average of the radiation from all directions at the test point. It can be measured with a black globe (often 150mm diameter) to give globe temperature at the test point, and corrected for any effects of air temperature and air velocity (see Parsons, 2014; ISO 7726, 1998). Pyranometers measure radiation directly and can be used to provide estimates of mean radiation temperature if averaged or plane radiant temperature (tpr) for directional radiation. If mean radiant temperature is considered in terms of an object or person then it is the average taking account of shape or posture so a weighted average is calculated. For solar radiation the principles of measurement are identical, but as solar radiation is a special case, and often dominant, it is often considered separately (Hodder and Parsons, 2007). Heat is transferred between objects in proportion to the differences in the fourth powers of their absolute surface temperatures. Rules of thumb are that if the globe temperature is less than air temperature or if surrounding surfaces are cooler than clothing or skin surface temperatures then that is an indication that there will be a net loss of heat from the body by radiation.

ISO 7726 (1998) recommends a sensor accuracy of ± 0.5 °C for temperatures above 0 °C and $\pm(0.5 - 0.01tr)$ °C for temperatures below 0 °C down to -40 °C with a response time as short as possible. Caution is provided in the use of globes that have long response times and rely on the accuracy of measures to correct for air temperature and air velocity. My experience is that if globes are understood and measures interpreted correctly then they have great practical utility.

Air velocity (va), including wind, is the rate at which air passes over the person. It is an important measure of cold stress as it determines wind chill. It is a three dimensional time varying phenomenon, usually considered as average wind speed over all directions. People can detect low levels of air movement so instruments require a measurement range of 0.05–10 ms^{-1} and above. Sensitive hot wire anemometers and kata thermometers can measure low levels of air movement and vein and cup anemometers can be used for higher levels such as those found in ventilation shafts and outdoors but also in cold stores and clean rooms. I have found the use of children's bubbles (video photographed with a mobile phone) to be useful for both estimating air speed and air velocity including direction and turbulence (bubbles can be messy and smoke and visible gases can also be useful but are sometimes intrusive). For cold stress measurement, ISO 7726 (1998) recommends a required accuracy of $\pm|0.1+0.05$val$|$ ms^{-1}.

Humidity is the amount of water vapour in the air and affects heat transfer from moist or wet skin and wet clothing. The amount of water the air can hold reduces with temperature so cold environments hold relatively little water vapour. The pressure exerted by the water vapour in the air is called the partial vapour pressure (pa) and the maximum water vapour that the air can hold at that temperature is termed 'the saturated vapour pressure' (ps,a). The relative humidity is pa/ps,a often expressed as a percentage. Very cold air is essentially dry air and heat and vapour is lost to the air from moist or wet skin at a saturated vapour pressure for skin temperature. For temperatures below 0 °C, freezing occurs so the heat transfer 'dynamic' changes. A wet bulb of the whirling hygrometer provides limited use for cold temperatures and dew point and condenser systems provide some information. For cold stress measurement, ISO 7726 (1998) recommends a required accuracy of ±0.15 kPa and a range down to 0.5 kPa.

Clothing and metabolic heat production in cold environments are considered in Chapters 4, 5 and 6. For estimated values the actual clothing and activity should be recorded and estimated values obtained from ISO 9920 (2007) and ISO 8996 (2004) (see Figures 4.1 and 5.1 and Parsons, 2014). The opportunity for people to influence the cold stress they receive can be estimated using a checklist (ability to adjust clothing insulation, increase activity, move around, take rest breaks in a warm environment, etc.). It can also be estimated using observation (including photographic) techniques. Actual behavioural responses will demonstrate which available opportunities are taken up (and by whom) and if any inappropriate behaviours occur.

MEASUREMENT OF COLD STRAIN

The response of a person to cold stress is 'cold strain' and its measurement includes physiological, subjective and behavioural measures. ISO 9886 (2004) describes physiological measures for assessing thermal strain. For cold stress, heart rate is a general index of strain and is a simple and convenient measure.

Internal body temperature indicates whole-body heat balance or if there is a tendency towards hypothermia. It can be measured at a number of sites on the body (e.g. rectal, oral and aural) and the site chosen will influence interpretation and precautions needed to achieve valid measures. Infra-red typmpanic measurement is most convenient but if the whole instrument becomes cold it may affect accuracy (as electronics cool) and cool air on skin and in the ear canal will influence results. In practical evaluation, taking the value at the beginning and end of a session in controlled conditions may be preferred.

Skin temperature is of great importance in assessing cold strain. To obtain mean skin temperature a number of sites are required across the body as, in the cold, skin temperature varies, and it is often impractical to measure at all sites. It is particularly useful for the evaluation of clothing. Measurement of extremity temperature (fingers, toes, ears, etc.) is useful for assessing the effects of cold on performance and possible injury. Little finger temperature and great toe temperature are often used to monitor health. Sweat loss is usually determined by weighing (before and after exposure) but is generally used in hot conditions. For exercise in the cold it may be best to observe sweating and consider chill during rest as a risk factor. Blood pressure is not a usual measure but may be useful especially in changing environments and in vulnerable people.

Subjective measures are very important in environmental surveys. It is useful to show a diagrammatic outline of a person with areas indicated and ask people to rate on relevant scales their feeling overall and at the indicated body sites. Often a sub-set of overall body and hand and foot sensations are used. Single sheet questionnaires or computer-based systems are useful.

ISO 28802 (2012) provides examples of subjective scales for sensation, discomfort, stickiness, preference, acceptability, satisfaction, draughtiness and dryness. The sensation scale is 5, extremely hot; 4, very hot; 3, hot; 2, warm; 1, slightly warm; 0, neutral; −1, slightly cool; −2, cool; −3, cold; −4, very cold; −5, extremely cold. In the region of comfort, the middle 7 terms are used (ISO 7730, 2005) and for cold conditions a truncated version from 1 or 0 to −5, for example, terms are used. For the discomfort or 'uncomfortable' scale, 4 points are used : 1, not uncomfortable; 2, slightly uncomfortable; 3, uncomfortable; 4, very uncomfortable, and similar for stickiness, draughtiness and dryness sometimes extended to '5, extremely …'. Acceptability and satisfaction are often bipolar scales (yes or no). Depending upon the aims of the survey specific questions can be added, about the effectiveness of clothing for example. A final set of open ended questions are important to collect opinions and avoid important missing information. Interview questionnaires are useful if possible, but it should be remembered that all surveys will be conducted in a political, social and psychological context. Discourse analysis and qualitative methods provide potential for improving environmental surveys but methods are in their infancy for use in this context.

Taken over space and time and for a representative sample of occupants of the cold environment (preferably while they are in it – how do YOU feel NOW?) subjective methods complement physical measurements and provide data suitable for presentation of outcomes as well as parametric (if continuous scales are used and assumptions allow) or non-parametric statistical analysis.

Behavioural measures are important as they need not interfere with the people or work and provide an indication of how a person responds when the environment begins to become unacceptable. Care must be taken in interpretation as cold may not be the only driver for behaviour. Structured (expected opportunities) and unstructured observation, often involving photographic techniques, are useful to collect data. Distraction (time off task) can be observed and quantified using these methods. Focus groups can also provide invaluable information if they are allowed (Van hoof, 2019).

ANALYSIS AND INTERPRETATION

The measurements of cold stress can be presented on a plan of the environment assessed as well as analysed to predict the likely response of people in the cold environment. ISO 11079 (2007) will provide an indication of the clothing required for health and comfort in the environment as well as guidance on wind chill, safe exposure times and limiting values for health and safety (see chapter 8). The 'HSDC' method described in chapter 10 will give an indication of likely loss in performance and the PMV/PPD method provided in ISO 7730 (2005) gives an indication of thermal discomfort and dissatisfaction in moderately cold environments. ISO 13732-3 (2005) will provide guidance on the consequences of bare skin touching cold surfaces (see Chapter 14).

CONCLUSION AND RECOMMENDATIONS

The conclusion of the survey will usually be in report form with a presentation to those who commissioned the work. A final conclusion must relate directly to the stated aims. Recommendations will depend upon the context and the opportunities to ensure an optimum or acceptable environment. Reduction of cold stress is a prime objective. This can be done by work re-organisation and providing behavioural opportunities. In an ice cream cold store that maintained an even air temperature of around −28 °C by circulating air, it was possible to switch off fans while people worked in that area and switch them on again when work was completed. Advice on clothing design and insulation can be provided from the calculation of IREQ and a task analysis (Shepherd and Stammers, 2005). This will require discussion with those involved and inform the final design of work systems.

CASE STUDY OF A COOK-CHILL HOSPITAL PLATING AREA

Parsons (1998) describes an assessment of a hospital plating area where, for reasons of hygiene, food delivery systems had moved from traditional (hot) kitchens to workers placing pre-prepared cold food (delivered in refrigerated lorries) to plates on a conveyer belt in a clean room held at 2 °C. The plates were then placed in trolley ovens for delivery to wards and patients so that food was at appropriate temperatures for consumption but at minimum time for contamination. Over all shifts, 3 men and 38 women performed the tasks. The health authority required advice on the actual cold stress in the chilled environment, advice on thermal comfort, welfare, clothing and any other areas relevant to staff working in the environment and to gain subjective assessments and opinions of staff working in the environment.

Only one day was allocated for this preliminary survey. Observation and enquiry revealed that the tasks in the plating area were cleaning the area (often by hand with soap and water), delivering and removing trolleys and food, arranging trays and serving the food. The staff wore extensive clothing, including thermal underwear, to an estimated insulation value of 2.48 clo inside the plating area and 2.0 clo outside of the plating area in 'normal' temperatures.

Air temperature, radiant temperature, humidity and air velocity were measured at all eight workplaces in the plating area. Equipment was calibrated before and after the measurements to ensure no drift. A 150mm diameter black globe provided globe temperature, thermistors provided air temperature, cross calibrated with a whirling hygrometer that also provided wet bulb temperature and hence humidity. Air velocity was measured with a hot wire anemometer and children's bubbles also showed the clean-room vertical ventilation system sucking air to the ceiling for exhaust. The room was empty as it was considered overly sensitive a topic for the experimenter to have direct contact with staff (an indication that there had been possible problems). The engineer was instructed to ensure that conditions were identical to workplace conditions. Measurements were made at ankle, chest and head height at each workplace. The metabolic rate was estimated as 70 Wm^{-2}.

Air temperature had been maintained between 1 and 3 °C throughout the 110 min assessment. Fans ensured a good air mix with gusts between 0.2 and 1.2 ms^{-1}. Globe temperature was similar to air temperature so it was assumed that there were minimal radiation effects and wet bulb and dry bulb temperatures were similar at around 2 °C so humidity was not of concern. ISO 11079 (2007) suggested that for the worst case of 1.6 °C air temperature and 1.2 ms^{-1} air velocity, with a basic clothing insulation value of 2.3 clo, a comfortable exposure time would be 1 hour and 20 minutes and if the fans were turned off for work a comfortable exposure time would be 1 hour and 40 minutes. The wind chill index indicated that the conditions were cold with fans switched on but no cold injuries could be expected. It was also within ACGIH guidelines.

A single sheet questionnaire, including a schematic diagram of a person showing areas of the body, thermal sensation, preference and satisfaction scales as well as an open catchall question, was administered by the health manager. It would have been preferable for the ergonomist to have administered the questionnaire but this was not possible or allowed. Eighteen questionnaires were completed. Information obtained included that in the plating area some staff had cold hands, nose and ears. Almost all respondents did not like the fans and almost all were too hot outside of the plating area and did not find changing clothing convenient.

A full report included recommendations. A starting clothing insulation of 2.5 clo, easily put on and off, with selection of individual garments by staff for extremities, should be provided. This should avoid thermal underwear and include an attractive but effective hat to cover hair net and ears. Fans should be turned off during the work period. Particular attention should be given to avoid wet clothing and skin when cleaning the area.

As with many consultancy cases, I was not informed about what happened after that!

Appendix 1

VBA Code for the Thermoregulatory Loop in the Modified Stolwijk and Hardy Model

The thermoregulatory loop receives as inputs starting conditions and values that do not change with time. It uses those to calculate new values after a fixed time. The program then loops back with the new values as starting values and calculates further values over the next fixed period, hence simulating responses to the cold conditions over time. When the time to view outputs is reached, it is printed to the spreadsheet (see Figure 9.1 and Chapter 9) up to the final exposure time when the program concludes. Symbols used in the code are described in Chapter 9.

Fixed values include environmental conditions, thermal properties and controller values. Modifications to this version of the model can allow them to change, for changing environmental conditions over exposure time for example. Starting values are represented as values for thermal comfort and these are changed with time as heat flows to and from the body.

KK is the spreadsheet row number for outputs and starts at 20 (Figure 9.1). 300 is a label for return to allow the loop

```
KK = 20
300
```

The first step is to determine the mean temperature of clothing (TCL(I)) and the radiant heat transfer coefficient (HR(I)) for each of the segments. As one is determined in a calculation involving the other, this is done by iteration as follows:

```
For I = 1 To 6
TCL(I) = 0
250 TCLOLD = TCL(I)
HR(I) = 4 * 5.67 * 10 ^ −8 * ((((TCL(I) + TR) / 2 + 273.2) ^ 3) *
    ARAD(I))
TCL(I) = ((1 / (CLO * 0.155)) * T(4 * I) + FACL * (HC(I) * TA + HR(I)
    * TR)) / ((1 / (CLO * 0.155)) + FACL * (HC(I) + HR(I)))
If Abs(TCL(I) − TCLOLD) > 0.01 Then GoTo 250
Next I
```

Adding heat transfer coefficients gives total heat transfer coefficient (H(I))

```
For I = 1 To 6: H(I) = (HR(I) + HC(I)) * S(I): Next I
```

The temperatures are compared with the set-point temperatures to determine the drive for change and the thermo-receptor outputs. For this application we will assume that there is no effect of the rate of change of temperatures, Rate=0. If the temperatures are less than set point then the body is cold and the variable DRIVER is negative.

```
For N = 1 To 25:
DRIVER(N) = 0: RATE(N) = 0: WARM(N) = 0: COLD(N) = 0
DRIVER (N) = T(N) − TSET(N) + RATE(N) * F(N)
If (DRIVER (N) < 0) Then COLD(N) = −DRIVER (N)
If (DRIVER(N) > 0) Then WARM(N) = DRIVER(N)
Next N
```

The COLD (and WARM) values from the skin of each segment (K=4*I) are added to integrate the 'peripheral afferents', giving an overall warm or cold disposition.

```
WARMS = 0: COLDS = 0
For I = 1 To 6: K = 4 * I
WARMS = WARMS + WARM(K) * SKINR(I): COLDS = COLDS +
    COLD(K) * SKINR(I): Next I
```

Using the driver as the difference between 'actual' and desired (set-point) temperatures and the sensitivity to any difference determined by the controllers, the physiological responses can be determined (the 'efferent outflows').

The extent of sweating for example is determined by the difference between head core temperature and its set point, the net sums of cold and warm receptor responses and the multiple of those, all moderated by sensitivity controllers. Vasodilation, vasoconstriction and shivering follow similar constructions.

$$SWEAT = CSW * DRIVER(1) + SSW * (WARMS - COLDS) + PSW$$
$$* DRIVER(1) * (WARMS - COLDS)$$
$$DILAT = CDIL * DRIVER(1) + SDIL * (WARMS - COLDS) + PDIL$$
$$* WARM(1) * WARMS$$
$$STRIC = -CCON * DRIVER(1) - SCON * (WARMS - COLDS) +$$
$$PCON * COLD(1) * COLDS$$
$$CHILL = (CCHIL * DRIVER(1) + SCHIL * (WARMS - COLDS))$$
$$*PCHIL * (WARMS - COLDS)$$

If (SWEAT <= 0) Then SWEAT = 0
If (DILAT <= 0) Then DILAT = 0
If (STRIC <= 0) Then STRIC = 0
If (CHILL <= 0) Then CHILL = 0

To prevent shivering in hot conditions the following is used.

If (COLD(1) = 0) Then If (COLDS = 0) Then CHILL = 0

Having determined effector output we must allocate it to the appropriate compartments (N) in each of the six segments. BULL determines the sensitivity of sweat gland response.

BULL = 4
For I = 1 To 6
N = 4 * I - 3
Q(N) = QB(N): BF(N) = BFB(N): E(N) = EB(N)
Q(N + 1) = QB(N + 1) + WORKM(I) * WORKI + CHILM(I) * CHILL
E(N + 1) = 0: BF(N + 1) = BFB(N + 1) + Q(N + 1) - QB(N + 1)
Q(N + 2) = QB(N + 2): E(N + 2) = 0: BF(N + 2) = BFB(N + 2)
Q(N + 3) = QB(N + 3): E(N + 3) = EB(N + 3) + SKINS(I) * SWEAT *
    2 ^ ((T(N + 3)) - TSET(N + 3)) / BULL
BF(N + 3) = (BFB(N + 3) + SKINV(I) * DILAT) / (1 + SKINC(I) * STRIC)

Calculate maximum evaporative capacity of the environment, EMAX(I), using Antoine's equation for Saturated vapour pressures for skin temperatures (N+3 =SVPTN3) and air temperature (SVPTA), the Lewis relation and evaporative resistance of clothing.

SVPTN3 = 0.13322 * (Exp(18.6686 − 4030.183 / (T(N + 3) + 235)))
SVPTA = 0.13322 * (Exp(18.6686 − 4030.183 / (TA + 235)))
LR(I) = 15.1512 * (TCL(I) + 273.15) / 273.15
RT(I) = Iecl + 1 / (FACL * LR(I) * HC(I))
EMAX(I) = ((1 / RT(I)) * (SVPTN3 − RH * SVPTA)) * S(I)
If (E(N + 3) > EMAX(I)) Then E(N + 3) = EMAX(I)
Next I

The dry heat exchange between the skin and the environment for each segment is the sum of radiant and convective heat exchanges (DRY(I)).

For I = 1 To 6: DRY(I) = (FACL * (HC(I) * (TCL(I) − TA) + HR(I) *
(TCL(I) − TR))) * S(I) : Next I

Heat exchange between blood and all compartments and between adjacent compartments is related to heat capacity and temperature difference.

For K = 1 To 24 : BC(K) = BF(K) * (T(K) − T(25)): TD(K) = TC(K) *
(T(K) − T(K + 1)): Next K

Heat flow into or out of each of the 24 compartments is related to metabolic heat production, minus heat by evaporation, blood flow and conduction apart from at the skin (HF(K+3)) where it includes heat loss to the environment (DRY(I)).

For I = 1 To 6: K = 4 * I − 3
HF(K) = Q(K) − E(K) − BC(K) − TD(K)
HF(K + 1) = Q(K + 1) − BC(K + 1) + TD(K) − TD(K + 1)
HF(K + 2) = Q(K + 2) − BC(K + 2) + TD(K + 1) − TD(K + 2)
HF(K + 3) = Q(K + 3) − BC(K + 3) − E(K + 3) + TD(K + 2) − DRY(I)
Next I

Add heat transfer due to blood flows across compartments, subtract heat losses from blood flow due to breathing, increment timer, and calculate new temperatures for compartments including blood.

HF(25) = 0: For K = 1 To 24: HF(25) = HF(25) + BC(K): Next K
HF(25) = HF(25) − 0.08 * WORKI
'Determine optimum integration step and calculate new temperatures
NTIME = 1 :DTIME = 1 / 60
For K = 1 To 25 : F(K) = HF(K) / C(K) : U = Abs(F(K))
If (U * DTIME > 0.1) Then DTIME = 0.1 / U : Next K

```
For K = 1 To 25 : T(K) = T(K) + F(K) * DTIME ; Next K
'Increment timer
RTIME = RTIME + DTIME : MTIME = 60 * RTIME
'when appropriate output results
If (MTIME >= KTIME) Then GoTo 700
GoTo 300
```

That is for each of the 25 compartments the new temperature is the old temperature plus the heat flow HF(K) in Watts divided by the heat capacitance of the compartment in W h °C^{-1} multiplied by DTIME in hours to give units of temperature in °C. In effect we add or subtract a temperature increment to each compartment, depending upon heat flow and thermal capacity.

The new temperatures and other values for the first time increment are then determined so we can go back to the beginning with the new values to calculate values for the second time increment and so on. This can be a loop that pauses to output results at the specified times (goto label 700) (see interface in Figure 9.1).

OUTPUT OF RESULTS

Results to output depend upon what is relevant to the application of interest. For human response to the cold, internal body temperature (brain) head core, T(1), will indicate overall heat loss if any and possible hypothermia. Mean skin (TS) and local hand T(16) and foot T(24) temperatures will indicate discomfort and possible local injury as well as possible effects on manual dexterity. Skin wettedness will indicate the state of the skin and metabolic heat will indicate shivering. Dry and Evaporative heat flows will indicate the dynamic nature of the interaction (Figure 9.1).

The above and additional factors can be calculated and presented to aid in interpretation. Cardiac output, CO, heat production, HP, insensible (evaporative) heat loss, EV, total dry heat transfer, DRYT, evaporative heat loss from the skin, ESK, skin wettedness, PWET, mean skin temperature, TS, mean body temperature, TB and total heat flow and skin blood flow, SBF are all set to zero for later calculation.

```
700 CO = 0: HP = 0: EV = 0: DRYT = 0: ESK = 0: PWET = 0: TS = 0:
    TB = 0: HFLOW = 0: SBF = 0
For N = 1 To 24 : CO = CO + BF(N) / 60: HP = HP + Q(N): EV = EV +
    E(N): Next N
```

MR = HP + WK: EV = EV + 0.08 * WORKI:
For I = 1 To 6: DRYT = DRYT + DRY(I): ESK = ESK + E(4 * I)
PWET = PWET + (E(4 * I) / EMAX(I)) * (S(I) / SA): SBF = SBF +
   BF(4 * I) / 60
TS = TS + T(4 * I) * C(4 * I) / 3.9: Next I
For N = 1 To 25: TB = TB + T(N) * C(N) / 68.79: HFLOW = HFLOW +
   HF(N): Next N

Divide by surface area to calculate tissue conductance in Wm^{-2}

EV = EV / SA: DRYT = DRYT / SA: ESK = ESK / SA: HP = HP / SA:
   MR = MR / SA: HFLOW = HFLOW / SA
COND = (HP − (E(1) + E(5)) / SA − HFLOW) / (T(25) − TS)

Print reuired outputs to the spreadsheet

Cells(KK, 12).Select: ActiveCell.Value = ITIME: Cells(KK, 13).Select:
   ActiveCell.Value = T(1)
Cells(KK, 14).Select: ActiveCell.Value = TS: Cells(KK, 15).Select:
   ActiveCell.Value = T(16)
Cells(KK, 16).Select: ActiveCell.Value = T(24): Cells(KK, 17).Select:
   ActiveCell.Value = PWET
Cells(KK, 18).Select: ActiveCell.Value = MR: Cells(KK, 19).Select:
   ActiveCell.Value = DRYT
Cells(KK, 20).Select: ActiveCell.Value = EV: Cells(KK, 21).Select:
   ActiveCell.Value = ESK
KK = KK + 1 moves the output to the next row
KTIME = KTIME + ITIME
If (MTIME < JTIME) Then GoTo 300
End Sub

The computer code demonstrates the versatility of the computer model that can be developed for the particular requirements of interest. The fundamental code is provided in Stolwijk and Hardy (1977) based upon NASA report CR-1855 by J A J Stowijk, Washington DC, 1971. Further versions are presented in Haslam and Parsons (1989a). Fiala (1998) developed the model to provide more detailed simulation (more layers and body parts) and Neale (1998) considered finite difference and finite element approaches using ubiquitous engineering software developed for computer-aided design. A good starting point for the most recent consideration is to refer to Fiala et al (2012) who used a version of the Fiala (1998) model as part of the development of the UTCI (Universal Thermal Climate Index) for use in weather forecasting.

References

ACGIH, 2021, *Threshold Limit Values (TLVs) and Biological Exposure Indices (BEIs)*. Cincinnati, OH: ACGIH American Conference of Governmental Industrial Hygienists.

ACSM, 1984, Prevention of thermal injuries during distance running. *Medicine and Science in Sports and Exercise*, 16(5), 427–443.

Armstrong, L. E., 2000, *Performance in Extreme Environments*. Champaign, IL: Human Kinetics.

Ashcroft, F., 2000, *Life at the Extremes*. London: Harper Collins.

ASHRAE, 2009, Thermal comfort. In *ASHRAE Handbook of Fundamentals*. Atlanta, GA: ASHRAE, pp. 9-1–9-33.

ASTM, 2020, ASTM F3426-20 Standard Test Method for Measuring the Thermal Insulation of Clothing Items Using Heated Manikin Body Forms, Developed by Subcommittee: F23.60, Book of Standards Volume: 11.03, ASTM, Washington, USA.

Belding, H. S., and Hatch, T. F., 1955, Index for evaluating heat stress in terms of resulting physiological strain. *Heating, Piping, and Air Conditioning*, 27, 129–136.

Bouskill, L., and Parsons, K. C., 1996, Effectiveness of air cooling air conditioning unit at reducing thermal strain during heat stress. In Robertson, S. A. (Ed.), *Contemporary Ergonomics*, pp. 510–514. Cirencester: Taylor and Francis.

Bouskill, L. M., Havenith, G., Kuklane, K., Parsons, K. C., and Withey, W. R., 2002, Relationship between clothing ventilation and thermal insulation. *American Industrial Hygiene Association Journal*, 63(3), 263–268.

Brager, G. S., Paliaga, G., and de Dear, R. J., 2004, Operable windows, personal control and occupant comfort. *ASHRAE Transactions*, 110(2), 17–35.

Bröde, P., Fiala, D., Blazejczyk, K., Holmer, I., Jendritszky, G., Kampman, B., Tinz, B., and Havenith, G., 2012, Deriving the operational procedure for the Universal Thermal Climate Index (UTCI). *International Journal of Biometeorology*, 56(3), 481–494.

Brooks, C. J., 2001, *Survival in Cold Waters*. AMSR Report TP 13822, Marine Safety-Transport Canada.

Brooks, J., and Parsons, K. C., 1999, An ergonomics investigation into human thermal comfort using an automobile seat heated with encapsulated carbonized fabric (ECF). *Ergonomics*, 42(5), 661–673.

Burton, A. C., and Edholm, O. G., 1955, *Man in a Cold Environment*. London: Edward Arnold.

Cabanac, M., 1995, *Human Selective Brain Cooling*. Germany: Springer Verlag.

Castellani, J. W., Sawka, M. N., DeGroot, D. W., Young, A. J., Castellani, J. W., et al., 2010, Cold thermoregulatory responses following exertional fatigue. *Frontiers in Bioscience*, 2, 854–865.

Castellani, J. W., and Tipton, M. J., 2015, Cold stress effects on exposure tolerance and exercise performance. *Comprehensive Physiology*, 6(1), 443–469.

Castellani, J. W., and Young, A. J., 2016, Human physiological responses to cold exposure: Acute responses and acclimatization to prolonged exposure. *Autonomic Neuroscience*, 196, 63–74.

Castellani, J. W., Young, A. J., Ducharme, M. B., Giesbrecht, G. G., Glickman, E., and Sallis, R. E., 2006, American College of Sports Medicine position stand: Prevention of cold injuries during exercise. American College of Sports Medicine 2006. *Medicine & Science in Sports & Exercise*, 38(11), 2012–2029.

Clark, R. P., and Edholm, O. G., 1985, *Man and His Thermal Environment*, London: Edward Arnold.

Collins, K., 1983, *Hypothermia the Facts*. Oxford: Oxford University Press.

Collins, K. J., and Hoinville, E., 1980, Temperature requirements in old age. *Building Services Engineering Research and Technology*, 1(4), 165–172.

Conway, G. A., Husberg, B. J., and Lincoln, J., 1998, Cold as a risk factor in working life in the circumpolar region. In I. Holmer and K. Kuklane (Eds.), *Problems with Cold Work*, pp. 1–10. Solna: NIWL.

Coull, N., 2019, *Thermoregulatory Responses and Ageing: A Body Mapping Approach*. PhD thesis, Loughborough University, Loughborough, United Kingdom.

Crockford, 1991, Personal communication. BOHS Clothing Science Group, BOHS/ Ergonomics Society, United Kingdom.

Dannielson, U., 1998, Risk of frostbite. In Holmer, I., and Kuklane, K., (Eds.), *Problems with Cold Work*, pp. 133–135. Solna: NIWL.

de Dear, R., Xiong, J., Kim, J., Cao, B., 2020, A review of adaptive thermal comfort research since 1998. *Energy and Buildings*, 214, 109893.

DIN 33403–5, 1994, *Climate at Workplaces and Their Environment, Part 5. Ergonomic Design of Cold Workplaces*. Berlin: DIN.

Dodt, E., and Zotterman, Y., 1952, Mode of action of warm receptors. *Acta Physiologica Scandinavica*, 26, 345–357.

Douglas, C. G., 1911, A method for determining the total respiratory exchange in man. *Journal of Physiology*, 17–18.

Du bois, D., and Du bois, E. F., 1916, A formula to estimate surface area if height and weight are known. *Archives of Internal Medicine*, 17, 863.

Durnin, J. V. G. A., and Passmore, R., 1967, *Energy, Work and Leisure*. London: Heinemann Education.

Education and School Premises Regulations, 1999, *UK Statutory Instruments 1999*, No. 2. Legislation.gov.uk.

EN 511, 2006, *Protective Gloves against the Cold*. Brussels: CEN.

Enander, A., 1987, Effects of moderate cold on performance at psychomotor and cognitive tasks. *Ergonomics*, 30, 1431–1435.

Environment Canada, 2017, Wind chill-the chilling facts, permanent link to this catalogue record. http://publications.gc.ca/pub?id=9.700202&sl=0.

EPA, 2017, Physical stress management program. Chapter 7 of *Emergency Responder Health and Safety Manual*. United States Environmental Protection Agency. https://response.epa.gov/_healthsafetymanual/manual-index.htm.

Fanger, P. O., 1970, *Thermal Comfort*. Copenhagen: Danish Technical Press.

Faulkner, S. H., Ferguson, R. A., Gerrett, N., Hupperets, M., and Hodder, S. G., 2013, Reducing muscle temperature drop after warm-up improves sprint cycling performance. *Medicine & Science in Sports & Exercise*, 45(2), 359–365.

Fiala, D., 1998, *Dynamic Simulation of Human Heat Transfer and Thermal Comfort*. PhD thesis, De Montfort University, Leicester, United Kingdom.

Fiala, D., Havenith, G., Bröde, P., Kampmann, B., and Jendritzky, G., 2012, UTCI-Fiala multi-node model of human heat transfer and temperature regulation. *International Journal of Biometeorology*, 56(3), 429–441.

Filingeri, D., Fournet, D., Hodder, S., and Havenith, G., 2014, Body mapping of cutaneous wetness perception across the human torso during thermo-neutral and warm environmental exposures. *Journal of Applied Physiology*, 117(8), 887–889.

Fox, W. F., 1967, Human performance in the cold. *Human Factors*, 9, 203–220.

Gagge, A. P., Burton, A. C., and Bazett, H. C., 1941, A practical system of units for the description of the heat exchange of man with his thermal environment. *Science*, 94, 428–430.

Galway, T. J., and O'Sullivan, L. W., 2005, Computer aided ergonomics. In J. R. Wilson and N. Corlett (Eds.), *Evaluation of Human Work*, Chapter 28, pp. 743–766. London: Taylor and Frnacis.

Garg, A., Chaffin, D. B., and Herrin, G. D., 1978, Prediction of metabolic rates for manual materials handling jobs. *American Industrial Hygiene Association Journal*, 39(8), 661–674.

Gerrett, N., Ouzzahra, Y., Redortier, B., Voelcker, T., and Havenith, G., 2015, Female sensitivity to hot and cold during rest and exercise. *Physiology & Behavior*, 152, 11–19.

Givoni, B., and Goldman, R. F., 1971, Predicting metabolic energy cost. *Journal of Applied Physiology*, 30(3), 429–433.

Givoni, B., and Goldman, R. F., 1972, Predicting rectal temperature response to work, environment and clothing. *Journal of Applied Physiology*, 2(6), 812–822.

Gold, E, 1935, The effect of wind, temperature, humidity and sunshine on the loss of heat of a body at temperature 98°F. *Journal of Royal Meteorological Society*, 61, 313–343.

Golden, F., and Tipton, M., 2002, *Essentials of Sea Survival*. Leeds: Human Kinetics.

Golden, F. S., and Hervey, G. R., 1977, The mechanism of the after-drop following immersion hypothermia in pigs. *Journal of Physiology*, 272, 26–27.

Golden, F. S., and Hervey, G. R., 1981, The "after-drop" and death after rescue from immersion in cold water. In J. A. Adam (Ed.), *Hypothermia Ashore and Afloat*, pp. 37–56. Aberdeen: Aberdeen University Press.

Goldman, R. F., and Kampmann, B., 2007, *Handbook on Clothing*, 2nd edn. NATO Research Study Group 7 on Bio-Medical Research Aspects of Military Protective Clothing. http://www.lboro.ac.uk/microsites/lds/EEC/ICEE/textsearch/Handbook%20on%20 Clothing%20-%202nd%20Ed.pdf.

Goldman, R. F., 1994, Local finger insulation and its effect on cooling rate. In *Proceedings of the Sixth International Conference on Environmental Ergonomics*, p. 84, Motebello, Canada.

Gordon, K., Blondin, D. P., Friesen, B. J., Tingelstad, H. C., Kenny, G. P., and Haman, F., 2019, Seven days of cold acclimation substantially reduces shivering intensity and increases nonshivering thermogenesis in adult humans. *Journal of Applied Physiology*, 26(6), 1598–1606.

Gordon, R. G., 1974, *The Response of a Human Temperature Regulator Model in the Cold*. PhD thesis, University of California, Santa Barbara, CA, United States.

Griffin, M. J., 1990, *Handbook of Human Vibration*. London: Academic Press.

Hamlet, M. P., 1988, Human cold injuries. In K. B. Pandolf, M. N. Sawka and Gonzalez, R. R. (Eds.), *Human Performance Physiology and Environmental Medicine at Terrestrial Extremes*, pp. 435–466. Dubuque, IA: Brown and Benchmark.

Hamlet, M. P., 1997, Peripheral cold injury. In I. Holmer and K. Kuklane (Eds.), *Problems with Cold Work*, pp. 127–131. Solna: National Institute for Working Life.

Hanson, H. E., and Goldman, R. F., 1969, Cold injury in man: A review of its etiology and discussion of its prediction. *Military Medicine*, 134(11), 1307–1316.

Hardy, R. N., 1979, *Temperature and Animal Life*, 2nd ed. London: Edward Arnold.

Hashiguchi, N., Yue, A. E., Feng, A. E., and Toshihara, Y., 2010, Gender differences in thermal comfort and mental performance at different vertical air temperatures. *European Journal of Applied Physiology*, 109, 41–48.

Haslam, R. A., and Parsons, K. C., 1989a, *Models of Human Response to Hot and Cold Environments*. Human Modelling Group Final Report, Vols 1 & 2, APRE, Farnborough.

Haslam, R. A., and Parsons, K. C., 1989b, Computer-based models of human responses to the thermal environments - Are their predictions accurate enough for practical use? In J. B. Mercer (Ed.), *Thermal Physiology*, pp. 763–768. Amsterdam: Elsevier.

Hayes, P., 1989, A physiological basis of cold protection. In J. B. Mercer (Ed.), *Thermal Physiology*, pp. 45–61. Amsterdam: Elsevier.

Hayward, J. S., Eckerson, J. D. and Collis, M. L., 1975, Thermal balance and survival time prediction of man in cold water. *Canadian Journal of Physiology and Pharmacology*, 53, 21–32.

Hill, L., 1919, *The Science of Ventilation and Open Air Treatment. Part I*. Report for the Medical Research Council, London, No. 32.

HMSO, 1963, *Offices, Shops and Railway Premises Act 1963*. London: HMSO.

Hodder, S. G., and Parsons, K. C., 2007, The effects of solar radiation on thermal comfort. *International Journal of Biometeorology*, 51(3), 233–250.

Holmer, I., 1984, Required clothing insulation (IREQ) as an analytical index of cold stress. *ASHRAE Transactions*, 90(1), 116–128.

Horvath, S. M., and Freedman, A., 1947, The influence of cold upon the efficiency of man. *Journal of Aviation Medicine*, 18, 158–164.

ISO 7726, 1998, *ED 2, Ergonomics of the Thermal Environment—Instruments for Measuring Physical Quantities*. Geneva: International Organization for Standardization.

ISO 7730, 2005, *ED 3, Ergonomics of the Thermal Environment—Analytical Determination and Interpretation of Thermal Comfort Using Calculation of the PMV and PPD Indices and Local Thermal Comfort Criteria.* Geneva: International Organization for Standardization.

ISO 7933, 1989, *Hot Environments—Analytical Determination and Interpretation of Thermal Stress Using Calculation of Required Sweat Rate.* Geneva: International Organization for Standardization.

ISO 8996, 2004, *ED 2, Ergonomics of the Thermal Environment—Determination of Metabolic Rate.* Geneva: International Organization for Standardization.

ISO 9886, 2004, *Evaluation of Thermal Strain by Physiological Measurements.* Geneva: International Organization for Standardization.

ISO 9920, 2007, *ED 2, Estimation of Thermal Insulation and Water Vapour Resistance of a Clothing Ensemble (See also Amended Version 2009).* Geneva: International Organization for Standardization.

ISO 11079, 2007, *ED 1, Ergonomics of the Thermal Environment—Determination and Interpretation of Cold Stress When Using Required Clothing Insulation (IREQ) and Local Cooling Effects.* Geneva: International Organization for Standardization.

ISO 11092, 1993, *Textiles —Physiological Effects — Measurement of Thermal and Water Vapour Resistance under Steady-State Conditions (Sweating Guarded-Hotplate Test).* Geneva: International Organization for Standardization.

ISO 12894, 2001, *Ergonomics of the Thermal Environment - Medical Supervision of Individuals Exposed to Extreme Hot or Cold Environments.* Geneva: International Organization for Standardization.

ISO 13732-3, 2005, *ED 1, Ergonomics of the Thermal Environment—Methods for the Assessment of Human Responses to Contact with Surfaces—Part 3: Cold Surfaces.* Geneva: International Organization for Standardization.

ISO 15743, 2008, *Ergonomics of the Thermal Environment - Cold Work Places - Risk Assessment and Management.* Geneva: International Organization for Standardization.

ISO 22411, 2021, *Ergonomics Data for Use in the Application of ISOIEC Guide 71, 2014.* Geneva: International Organization for Standardization.

ISO 28802, 2012, *Ergonomics of the Physical Environment – Assessment of Environments by Means of an Environmental Survey Involving Physical Measurements of the Environment and Subjective Responses of People.* Geneva: International Organization for Standardization.

ISO 28803, 2012, *Ergonomics of the Physical Environment – Application of International Standards to Physical Environments for People with Special Requirements.* Geneva: International Organization for Standardization.

ISO CD 23454–1, 2020, *Human Performance in Physical Environments: Part 1. A Performance Framework.* Geneva: International Organization for Standardization.

ISO NWI 23454–2. *Human Performance in Physical Environments: Part 2-Measures and Methods for Assessing the Effects of the Physical Environment on Human Performance.* Geneva: International Organization for Standardization.

Jelen, W., and Syrstad, B., 2010, *VBA and Macros Microsoft Excel 2010*. Indianapolis, IN: Que Publishing.

Keatinge, W. R., and Donaldson, G. C., 1998, Differences in cold exposures associated with excess winter mortality. In Holmer, I., and Kuklane, K. (Eds.), *Problems with Cold Work*, pp. 210–215. Solna: NIWL.

Keatinge, W. R., 1969, *Survival in Cold Water*. Oxford: Blackwell Scientific.

Kenney, L. W., Wilmore, J. H., and Costill, D. L., 2012, *Physiology in Sport and Exercise*, 5th ed. Champaign, IL: Human Kinetics.

Kerslake, D. M., 1972, *The Stress of Hot Environment*. Cambridge: Cambridge University Press.

LeBalnc, J., 1975, *Man in the Cold*. Springfield, IL: C. Thomas.

Legg, S. J., and Pateman, C. M., 1984, A physiological study of the repetitive lifting capabilities of healthy young males. *Ergonomics*, 27(3), 259–272.

Lewis, W. K., 1922, The evaporation of a liquid into gas. *ASME Transactions*, 44, 325–335

Liddell, D. K., 1963, Estimation of energy expenditure from expired air. *Journal of Applied Physiology*, 18, 25–29.

Mackworth, N. H., 1953, Finger numbness in very cold winds. *Journal of Applied Physiology*, 5, 533–543.

McCullough, E. A., Jones, B. W., and Huck, J., 1985, A comprehensive database for estimating clothing insulation. *ASHRAE Transactions*, 91(2A), 29–47.

McCullough, E. A., Jones, B. W., and Tamura, T., 1989, A database for determining the evaporative resistance of clothing. *ASHRAE Transactions*, 95, 316–328.

McIntyre, D. A., 1980, *Indoor Climate*. London: Applied Science.

Meese, G. B., Kok, R., Lewis, M. I., and Wyan, D. P., 1984, A laboratory study of the effects of moderate thermal stress on the performance of factory workers. *Ergonomics*, 27(1), 19–43.

Militky, J., and Kremenakova, D., 2008, Thermal conductivity of wool/pet weaves. Paper MJ2. In *Proceedings of the Sixth International Conference on Heat Transfer, Fluid Mechanics and Thermodynamics. HEFAT 200*, South Africa.

Morrissey, S. J., and Liou, Y. H., 1984, Metabolic cost of load carriage with different container sizes. *Ergonomics*, 27(8), 847–853.

Murgatroyd, P. R., Shetty, P. S., and Prentice, A. M., 1993, Techniques for the measurement of human energy expenditure: A practical guide. *International Journal of Obesity*, 17, 549–568.

Neale, M. S., 1998, *Development and Application of a Clothed Thermoregulatory Model*. PhD thesis, Loughborough University, Loughborough, United Kingdom.

Newburgh, L. H. (Ed.), 1957, *The Physiology of Temperature Regulation and the Science of Clothing*. New York: Hafner, pp. 374–388.

Nishi, Y., and Gagge, A. P., 1977, Effective temperature scale useful for hypo and hyperbaric environments. *Aviation, Space, and Environmental Medicine*, 48, 97–107.

Oakley, E. H. N., 1990, A new mathematical model of finger cooling used to predict the effects of windchill and subsequent liability to freezing cold injury. In *Proceedings of International Conference in Environmental Ergonomics - IV*, Austin, TX, United States.

Oakley, E. H. N., and Lloyd, C. J., 1990, Investigations into the pathophysiology of mild cold injury in human subjects. In *Proceedings of International Conference in Environmental Ergonomics - IV*, Austin, TX, United States.

Oakley, E. H. N., and Petherbridge, R. J., 1997, *The Prediction of Survival during Cold Immersion. Results from the National Immersion Incident Survey*. Institute of Naval Medicine. Report No. 97011. Alverstoke, United Kingdom.

Olesen, B. W., 1985, *Local Thermal Discomfort, Technical Review No. 1*. Copenhagen: Bruel and Kjaer.

Osczevski, R., and Bluestein, M., 2001, *A New Wind Chill Index. In Wind Chill Science and Equations*. World Wide Web report by Environment Canada and DCIEM, Toronto, ON, Canada.

Pandolf, K. B., Givoni, B., and Goldman, R. F., 1977, Predicting energy expenditure with loads while standing or walking very slowly. *Journal of Applied Physiology*, 43(4), 577–581.

Parsons, K. C., 1991, User performance tests for determining the thermal properties of clothing. In Queinnec, Y., and Daniellou, F. (Eds.), *Designing for Everyone*. In *Proceedings of the 11th Congress of the International Ergonomics Association*. Paris: Taylor and Francis.

Parsons, K. C., 1992, The thermal audit, In Lovesey, E. J. (Ed.), *Contemporary Ergonomics*, pp. 85–90. London: Taylor and Francis.

Parsons, K. C., 1998, Case study of cold work in a hospital plating area. In I. Holmer and K. Kuklane (Eds.), *Problems with Cold Work*, pp. 66–68. Solna: NIWL.

Parsons, K. C., 2005a, The environmental ergonomics survey. In Wilson, J. and Corlett, N. (Eds.), *Evaluation of Human Work*, 3rd ed., Chapter 22. New York: Taylor and Francis.

Parsons, K. C., 2005b, Ergonomics assessment of thermal environments. In Wilson, J. and Corlett, N. (Eds.), *Evaluation of Human Work*, 3rd ed., Chapter 23, New York: Taylor and Francis.

Parsons, K. C., 2014, *Human Thermal Environments*, 3rd ed. New York: Taylor and Francis.

Parsons, K. C., 2018, ISO standards on physical environments for worker performance and productivity. *Industrial Health*, 56(2), 93–95.

Parsons, K. C., 2019, *Human Heat Stress*. Boca Raton, FL: CRC Press.

Parsons, K. C., 2020, *Human Thermal Comfort*. Boca Raton, FL: CRC Press.

Parsons, K. C., and Bishop, D., 1991, A data base model of human responses to thermal environments. In Lovesey, E. J. (Ed.), *Contemporary Ergonomics*, pp. 444–449. London: Taylor and Francis.

Parsons, K. C., and Egerton, D. W., 1985, The effect of glove design on manual dexterity in neutral and cold conditions. In D. J. Oborne (Ed.), *Contemporary Ergonomics*, pp. 203–209. London: Taylor and Francis.

Parsons, K. C., and Haslam, R. A., 1984, Evaluation of a model of human thermoregulation and its possible ergonomics applications. In E. D. MeGaw (Ed.), *Contemporary Ergonomics*, pp. 134–141. London: Taylor and Francis.

Parsons, K. C., and Hamley, E. L., 1989, Practical methods for the estimation of human metabolic heat production. In J. B. Mercer (Ed.), *Thermal Physiology*, pp. 777–781. Amsterdam: Elsevier.

Parsons, K. C., and Webb, L. H., 1999, *Thermal Comfort Design Conditions for Indoor Environments Occupied by People with Physical Disabilities*. Final Report to EPSRC Research Grant GRK71295, Loughborough University, United Kingdom.

Parsons, K. C., and Webb, L. H., 2013, Thermal environments for people with physical disabilities. In Cotter, J. D., Lucas, S. J. E., and Mundel, T. (Eds.), *Environmental Ergonomics*, p. 206. Conference proceedings, ICEE Queenstown, New Zealand.

Payne, R. B., 1959, Tracking proficiency as a function of thermal balance. *Journal of Applied Physiology*, 14, 387–389.

Pierce, F. T., and Rees, W. H., 1946, The transmission of heat through textile fabrics, Part II. *Journal of the Textile Institute*, 37, 181–204.

Potter, A. W., Looney, D. P., Santee, W. R., Gonzalez, J. A., Welles, A. P., Srinivasan, S., Castellani, M. P., Rioux, T. P., Hansen, E. O., and Xiaojiang, X., 2020, Validation of new method for predicting human skin temperatures during cold exposure. The cold weather ensemble decision aid (CoWEDA). *Informatics in Medicine Unlocked*, 2020, 100301.

Pozos, R., and Danzl, M. D., 2001, Human physiological reactions to cold stress and hypothermia. In Pandolf, K. B., and Burr, R. E. (Eds.), *Medical Aspects of Harsh Environments*. Vol. 1. Office of the Surgeon General of the United States Army.

Provins, K. A., and Morton, R., 1960, Tactile discrimination and skin temperature. *Journal of Applied Physiology*, 15, 155–160.

Ramanathan, N. L., 1964, A new weighting system for mean surface temperature of the human body. *Journal of Applied Physiology*, 19, 531–553.

Randle, I. P. M., 1987, Predicting the metabolic cost of intermittent load carriage in the arms. In E. D. Megaw (Ed.), *Contemporary Ergonomics*, pp. 286–291. London: Taylor and Francis.

Randle, I. P. M., Legge, S. J., and Stubbs, D. A., 1989, Task-based prediction models for intermittent load carriage. In E. D. Megaw (Ed.), *Contemporary Ergonomics*, pp. 380–385. London: Taylor and Francis.

Reegan Alicia, 2017, Regulating your body temperature with a spinal cord injury. https://ablethrive.com/life-skills/regulating-your-body-temperature-spinal-cord-injury.

Reinertsen, R. E., 1998, Occupational cold exposure in the offshore environment; development of test methods for protective clothing. In I. Holmer and K. Kuklane (Eds.), *Problems with Cold Work*, pp. 11–12. Solna: NIWL.

Renbourn, E. T., 1972, *Materials and Clothing in Health and Disease*. London: H K Lewis.

Ringuest, 1981, *A Statistical Model of the Controller Function of the Human Temperature Regulating System*. PhD thesis, Clemson University, United States.

Rivolier, J., Goldsmith, R., Lugg, D. J., and Taylor, A. J., 1988, *Man in the Antarctic*. London: Taylor and Francis.

Santee, W. R., and Berglund, L. A., 2001, Thermal properties of handwear at varying altitudes. *Aviation, Space, and Environmental Medicine*, 72, 576–578.

Santee, W. R., and Gonzalez, R. R., 1988, Characteristics of the thermal environment. In K. B. Pandolf, M. N. Sawka, and R. R. Gonzalez (Eds.), *Human Performance Physiology and Environmental Medicine at Terrestrial Extremes*, pp. 1–44. Dubuque, IA: Brown and Benchmark.

Santee, W. R., Potter, A. W., and Freidl, A. D., 2017, Talk to the hand: U.S. army biophysical testing. *Military Medicine*, 182(7): e1702–e1705.

Shepherd, A. S., and Stammers, R. B., 2005, Task analysis. In J. R. Wilson and E. N. Corlett (Eds.), *Evaluation of Human Work*, pp. 129–158, Chapter 6. London: Taylor and Francis.

Shitzer, A., and Tikuisis, P., 2012, Advances, shortcomings, and recommendations for wind chill estimation. *International Journal of Biometeorology*, 56(3), 495–503.

Siple, P. A., 1945, General principles governing selection of clothing for cold climates. *Proceedings of the American Philosophical Society*, 89(1), 200–234.

Siple, P. A., and Passel, C. F., 1945, Measurements of dry atmosphere cooling in subfreezing temperatures. *Proceedings of the American Philosophical Society*, 89(1), 177–199.

Slonim, A. D., 1952, *Fundamentals of the General Ecological Physiology of Mammals*. Moscow: Academic Press of USSR.

Spitzer, H., and Hettinger, T., 1976, *Caloricentafels. Tabellen voorhet omzetten van fysissche activiteiten in Calorisch Waarden*. Leuven: Acco.

Stolwijk, J. A. J., 1971, *A Mathematical Model of Physiological Temperature Regulation in Man*. Washington, DC: NASA. NASA Report No. 1855.

Stolwijk, J. A. J., and Hardy, J. D., 1966, Temperature regulation in man - A theoretical study. *PfluÈ ger Archives*, 291, 129–162.

Stolwijk, J. A. J., and Hardy, J. D., 1977, Control of body temperature. In *Handbook of Physiology, Section 9: Reaction to Environmental Agents*, pp. 45–68. Bethesda, MD: American Physiological Society.

Tochihara, Y., 1998, Work in artificially cold environments. In I. Holmer and K. Kuklane (Eds.), *Problems with Cold Work*, pp. 13–15. Solna: NIWL. San Diego Naval Health Research Center and San Diego State University, ISBN 0966695313.

Tochihara, Y., Kimura, Y., Yadoguchi, I. U., and Nomura, M., 1998, Thermal responses to air temperature before, during and after bathing. In J. A. Hodgdon, J. H. Heaney, and M. J. Buono (Eds.), *Environmental Ergonomics VIII*, pp. 309–313. San Diego, CA: Naval Health Research Center and San Diego State University.

Todd, S., 1988, *Can Performance of a Manual Task Be Predicted from Hand Skin Temperature in Cold Conditions?* Final Year Undergraduate Ergonomics Project, Loughborough University, United Kingdom.

Toner, M. M., and McArdle, W. D., 1988, Physiological adjustments of man to the cold. In K. B. Pandolf, M. N. Sawka, and R. R. Gonzalez (Eds.), *Human Performance Physiology and Environmental Medicine at Terrestrial Extremes*, pp. 361–400. Dubuque, IA: Brown and Benchmark.

Underwood, C. R., and Ward, E. J., 1966, The solar radiation area of man. *Ergonomics*, 10, 399–410.

Underwood, P., and Parsons, K. C., 2005, Discomfort caused by sitting next to a cold window. Simulated railway carriage at night. In *Contemporary Ergonomics*, pp. 23–238. Oxford, UK: Taylor and Francis.

Vogt, J. J., Candas, V., Libert, J. P. and Sagat, J. C., 1981, Required sweat rate as an index of thermal strain in industry. In K. Cena and J. A. Clark (Eds.), *Bioengineering Thermo-Physiology and Comfort*, pp. 99–110. Amsterdam: Elsevier.

Van Dill, A. M., Day, R., and Siple, P. A., 1957, Special problems of hands. In L. H. Newburgh (Ed.), *The Physiology of Temperature Regulation and the Science of Clothing*, pp. 374–388. New York: Hafner.

Van Hoof, J., Bennetts, H., Hansen, A., Kazak, K. J., and Soebarto, V., 2019, The living environment and thermal behaviours of older South Austrlians: A multifocus group study. *International Journal of Environmental Research and Public Health*, 16, 935.

Wadsworth, P. M., and Parsons, K. C., 1989, The design, development, evaluation and implementation of an expert system into an organisation. In *Proceedings of the 3rd International Conference on Human–Computer Interaction*. Amsterdam: Elsevier.

Webb, L. H., and Parsons, K. C., 1997, Thermal comfort requirements for people with physical disabilities. In *Proceedings of the BEPAC and EPSRC Mini Conference: Sustainable Buildings*, pp. 114–121. Oxford: Abingdon.

Webb, L. H., Parsons, K. C., and Hodder, S. G., 1999, Thermal comfort requirements, a study of people with multiple sclerosis. *ASHRAE Transactions*, 105, 648.

Weir, J. B. de V., 1949, New methods for calculating metabolic rate with special reference to protein metabolism. *Journal of Physiology*, 109, 1–9.

Whyndam, C. H. and Atkins, A. R., 1966, A study of temperature regulation in the human body with the aid of an analogue computer. In *Proceedings of the 3rd International Conference of Medical Electronics*, London.

Wilson, J., and Sharples, S., 2015, *Evaluation of Human Work*, 3rd ed. Oxford, UK: Taylor and Francis, Oxford, UK.

Wilson, O., and Goldman, R. F., 1970, Role of air temperature and wind in the time necessary for a finger to freeze. *Journal of Applied Physiology*, 29(5), 658–664.

Winslow, C. E. A., Herrington, L. P., and Gagge, A. P., 1936, A new method of partitional calorimetry. *American Journal of Physiology-Legacy Content*, 116, 669.

Wissler, E. H., 1961, Steady-state temperature distribution in man. *Journal of Applied Physiology*, 16, 734–740.

Wissler, E. H., 1984, Mathematical simulation of human thermal behavior using whole-body models. In Shitzer, A., and Eberhart, R. C. (Eds.), *Heat Transfer in Medicine and Biology*, Chapter 13, Vol. 1, pp. 325–374. New York: Plenum Press.

Woodcock, A. H., 1962, Moisture transfer in textile systems. *Textile Research Journal*, 8, 628–633.

Wyon, D. P., 2001, Thermal effects on performance. In Spengler, J. D., McCarthy, J. F., and Samet, J. M. (Eds.), *Indoor Air Quality Handbook*, pp. 16.1–16.14. New York: McGraw-Hill.

Yoshida, J. A., Banhidi, L., Polinezky, T., Kintses, G., Hachisu, H., Imai, H., Sato, K., and Nonaka, M., 1993, A study on thermal environment for physically handicapped persons. Results from Japanese – Hungarian joint experiment in 1990. *Journal of Thermal Biology*, 18, 363–375.

Young, A. J., 1988, Human adaptation to cold. In K. B. Pandolf, M. N. Sawka, and R. R. Gonzalez (Eds.), *Human Performance Physiology and Environmental Medicine at Terrestrial Extremes*, pp. 401–434. Dubuque, IA: Brown and Benchmark.

Zhou, G. H., Loveday, D. L., Taki, S. A. H., and Parsons, K. C., 2002, Measurement of the air flow and temperature fields around live subjects and the evaluation of human heat loss. In *Proceedings of Indoor Air Conference*, July 2002, Monterey, CA, United States.

Index